# Transmission of Electrical Power

## Dr. Hidaia Mahmood Mohamed Alassouli

Hidaia_alassouli@hotmail.com

# Introduction

This book includes my lecture notes for electrical power transmission course. The power transmission process, from generation to distribution is described and expressions for resistance, inductance and capacitance of high-voltage power transmission lines are developed used to determine the equivalent circuit of a three-phase transmission line.

**The book is divided to different learning outcomes**

Part 1- Describe the power transmission process, from generation to distribution.

Part 2- Develop expressions for resistance, inductance and capacitance of high-voltage power transmission lines and determine the equivalent circuit of a three-phase transmission line.

**Part 1: Describe the power transmission process, from generation to distribution.**

- Describe the components of an electrical power system.
- Identify types of power lines, standard voltages, and components of high-voltage transmission lines (HVTL).
- Describe the construction of a transmission line, galloping lines, corona effect, insulator pollution, and lightning strikes.
- Explain transmission system stability in regards to power transfer, power flow division, and transfer impedance.

**Part 2: Develop expressions for resistance, inductance and capacitance of high-voltage power transmission lines and determine the equivalent circuit of a three-phase transmission line.**

- List the types of conductors used in power transmission line.
- Develop the expression for the inductance and capacitance of a simple, single-phase, two wire transmission line composed of solid round conductors.
- Deduce the expression for the inductance and capacitance of a simple, single-phase composite (stranded) conductor line.
- Derive the expression for the inductance and capacitance of three-phase lines having symmetrically and asymmetrically spacing and for bundled conductors.
- Discuss the effect of earth on the capacitance of three-phase transmission lines.
- Derive the short transmission lines models and medium transmission lines models.

# EEL 2023

# •POWER GENERATION AND
# •TRANSMISSION

## LO 3
### Chapter 4: Basic Concepts of Electrical Power Transmission System

0

## Power System Parts

o A power system is made of the following parts:
- Generation
- Transmission
- Distribution
- Substation for voltage/current conversions

o Function of the Transmission system:
- The main function of the transmission system is to transmit bulk power to load centers and large industrial users.

1

# TRANSMISSION LINES

o Generators and loads are connected together through transmission lines transporting electric power from one place to another. Transmission lines must, therefore, take power from generators, transmit it to locations where it will be used, and then distribute it to individual consumers.

o The power capability of a transmission line is *proportional to the square of the voltage* on the line. Therefore, very high voltage levels are used to transmit power over long distances. Once the power reaches the area where it will be used, it is stepped down to lower voltages in distribution substations, and then delivered to customers through distribution lines.

o Power cables can be either *Overhead Lines* or *Underground Lines*

# TRANSMISSION LINES

o **The first components that we will consider are Transmission Lines.**

o **We start with Characteristics of Power Lines.**

o **Transmission lines have resistance, inductance and capacitance**

# Resistance

- The DC resistance of a conductor is given by

$$R_{DC} = \frac{\rho l}{A}$$

- Where $\rho$ is the specific resistance (resistivity) of the material, $l$ is the length, and $A$ the cross-sectional area.
- Therefore, the DC resistance per meter of the conductor is

$$r_{DC} = \frac{\rho}{A} \left[\frac{\Omega}{m}\right]$$

- To keep the resistance of the lines as low as possible they should have a large diameter. Resistance reduces with the square of the radius.

4

- However, resistance must be weighed against other factors, including the cost of the conductor cable itself and its weight that needs to be supported by the towers
- The resistivity of a conductor is a fundamental property of the material that the conductor is made from. It varies with both type and temperature of the material.
- The resistivity increases linearly with temperature over normal range of temperatures.

5

- AC resistance of a conductor is always higher than its DC resistance due to the *skin effect* forcing more current flow near the outer surface of the conductor. The higher the frequency of current, the more noticeable skin effect would be.
- At frequencies of our interest (50-60 Hz), however, skin effect is not very strong.
- Wire manufacturers usually supply tables of resistance per unit length at common frequencies (50 and 60 Hz). Therefore, the resistance can be determined from such tables.

## INDUCTANCE AND INDUCTIVE REACTANCE

- Note that while resistance of lines is critical with regard to line losses, it is less important with regard to power flow and stability. This is because the overall impedance of a line tends to be dominated in practice by its inductive reactance, to such an extent that zero resistance is assumed.
- Recall that inductance is based on magnetic flux lines linking a loop of wire. This notion extends to a straight wire, which can be considered an *infinitely large loop*, and the magnetic flux around the wire does link it. Since there is only a fraction of a turn in a straight line, this magnetic effect is quite weak.

# Line Inductance

- But inductance is cumulative on a per–unit-length basis, and with a conductor that extends over tens or hundreds of miles, it does eventually add up.

- Indeed, there are two contributions to line inductance: the *self-inductance*, which is just a property of the individual conductor, and the *mutual inductance*, which occurs between the conductors of the three different phases.

8

# Inductance

- In most of the practical situations, the inductance of the transmission line can be found from tables supplied by line developers.
- Analysis of inductance properties shows that:

1. The *greater the spacing between* the phases of a transmission line, the *greater the inductance* of the line. Since the phases of a high-voltage overhead transmission line must be spaced further apart to ensure proper insulation, *a high-voltage line* will have a *higher inductance* than a *low-voltage line*.
Since the spacing between lines in buried cables is very small, series inductance of cables is much smaller than the inductance of overhead lines.

9

# Inductance

- For three phase circuit whose three circular conductors are at the corners of equilateral triangle(Fig-B(i)) then the above formula for single phase case is applied here. In this case inductance per phase L is as below:

If the Denominator is renamed as Ds, then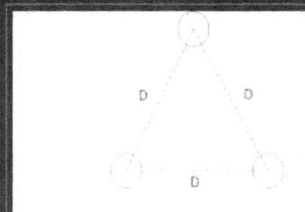

$$L = 2 * 10^{-7} \ln ( D / D_s )$$

Here Ds = r'

As already said r' is 0.7788 times the actual radius(r) of conductor.

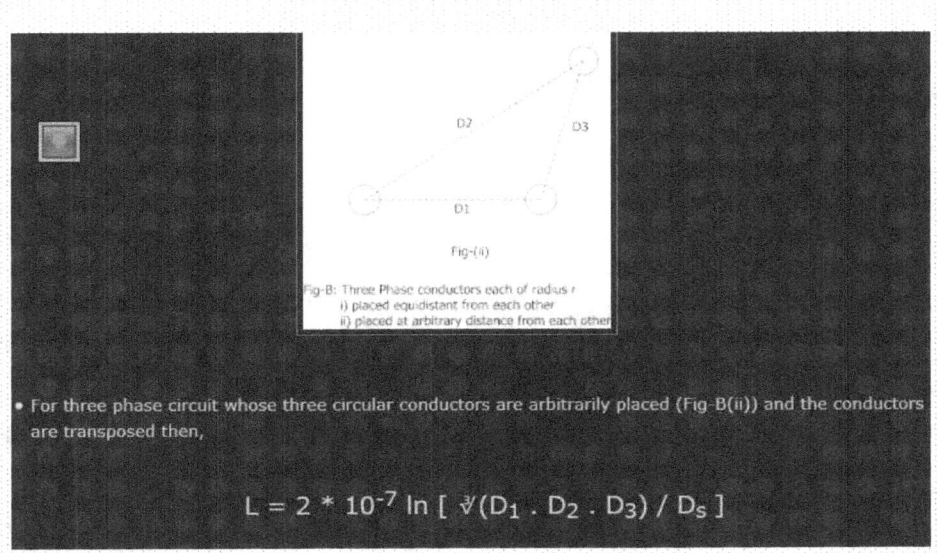

**Where**
D = Geometric Mean Distance (GMD)
$D_s$ = Geometric Mean Radius (GMR)

Fig-(ii)

Fig-B: Three Phase conductors each of radius r
i) placed equidistant from each other
ii) placed at arbitrary distance from each other

- For three phase circuit whose three circular conductors are arbitrarily placed (Fig-B(ii)) and the conductors are transposed then,

$$L = 2 * 10^{-7} \ln [ \sqrt[3]{(D_1 . D_2 . D_3)} / D_s ]$$

- 2. The greater the radius of the conductors in a transmission line, the lower the inductance of the line.

- In practical transmission lines, instead of using heavy and inflexible conductors of large radii, two and more conductors are bundled together to approximate a large diameter conductor.

- The more conductors included in the bundle, the better the approximation becomes. Bundles are often used in the high-voltage transmission lines.

# CAPACITANCE AND CAPACITIVE REACTANCE

- Transmission lines have capacitance, too. It is a bit easier to see how two lines travelling next to each other would vaguely resemble opposing plates with a gap in between.

- In fact, there is also capacitance between a conductor and the ground. Because the lines are small and the gap wide, the capacitance tends to be fairly small

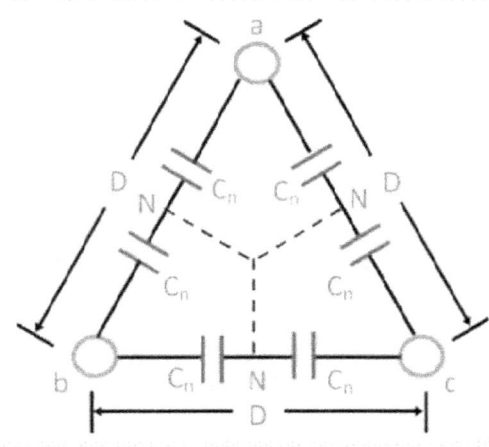

$$C_n = \frac{q_n}{V_{an}} = \frac{2\pi\varepsilon_0}{ln\frac{D}{r}} = \frac{1}{18\times10^9 ln\frac{D}{r}} F/m$$

○ Analysis of capacitive properties of lines shows that:

1. The greater the spacing between the phases of a transmission line, the lower the capacitance of the line. Since the phases of a high-voltage overhead transmission line must be spaced further apart to ensure proper insulation, a high-voltage line will have a lower capacitance than a low-voltage line. Since the spacing between lines in buried cables is very small, shunt capacitance of cables is much larger than the capacitance of overhead lines. Cable lines are normally used for short transmission lines (to min capacitance) in urban areas.

2-The greater the radius of the conductors in a transmission line, the higher the capacitance of the line.

○ Good transmission line is a compromise among the requirements for low series inductance, low shunt capacitance, and a large enough separation to provide insulation between the phases.

13

# Transmission Line Model

○ In describing transmission-line parameters, the inductance is generally considered to be in series and the capacitance in parallel.

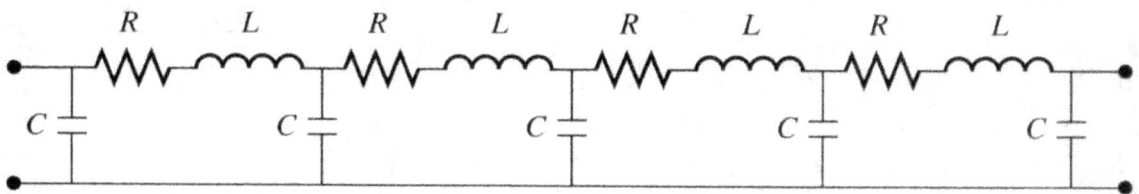

14

# SINGLE CORE OR BUNDLED

o Conductors on transmission lines—especially high-voltage, high-capacity lines—are sometimes bundled, meaning that what is electrically a single conductor is actually composed of two, three, or four wires a few inches apart, held together every so often with connectors known as conducting frames.

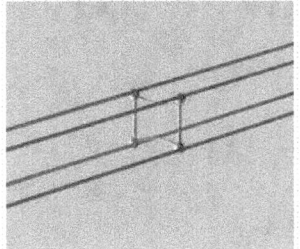 

# BUNDLING OF CONDUCTORS

There are several reasons for bundling conductors:

o Increasing heat dissipation as a result of increasing surface area,

o Reducing corona losses,

o Reducing inductance,

o Increasing the amount of current in the conductor due to reduced skin effect

# BUNDLING OF CONDUCTORS

- For 220 kV lines, two-conductor bundles are usually used, for 380 kV lines usually three or even four.

- A bundle conductor results in lower reactance, compared to a single conductor.

- As a disadvantage, the bundle conductors have higher wind loading.

17

# CORONA

- A corona is a process by which a current flows from an electrode with a high potential into a neutral fluid, usually air, by ionizing that fluid so as to create a region of ionized atoms around the electrode. The ions generated eventually pass charge to nearby areas of lower potential, or recombine to form neutral gas molecules.

- When the potential of the electric field is large enough at a point in the fluid, the fluid at that point ionizes and it becomes conductive. If a charged object has a sharp point, the air around that point will be at a much higher voltage gradient than elsewhere. Air near the electrode can become ionized (partially conductive), while regions more distant do not.

18

# CORONA

- One reason for bundling conductors is to reduce corona losses. The corona results from the electric field that surrounds the conductor at high voltage.

- Microscopic arcs occur between the conductor surface at high potential and ionized air molecules in the vicinity.

- The audible crackling sound around high-voltage a.c. equipment comes from the corona of tiny arcs that discharge into the air.

# CORONA

- http://www.youtube.com/watch?v=rLrP9mck7eM
- http://www.youtube.com/watch?v=ehQwzTxNZ9c

# CORONA

- Because the arcs are so small, they are not visible even at night. Yet there is a measurable energy loss associated with what is in fact a small electric current flowing to ground through the air. The power associated with this current is the corona loss.

- When the surface area of the conductor is increased, the electric potential or surface charge density is spread out more, reducing the electric field strength. This in turn reduces the formation of arcs, and thus reduces corona losses.

# MARKER BALLS

- http://www.youtube.com/watch?v=JAJu6nGYvh8#t=57
- http://www.youtube.com/watch?v=BQAl2r-PRpA

# PLANNING A TRANSMISSION SYSTEM

- **The decision to build a transmission system results from system planning studies to determine how best to meet the system requirements. The factors that need to be considered at the planning stage are discussed in the following sections.**

  - *Determination of Transmission Voltages*
  - *Deciding Between Overhead Lines and Underground Cables*

# DETERMINATION OF TRANSMISSION VOLTAGES

*System Voltages:*

- **There is no clear separation between distribution, sub- transmission, and transmission voltage levels. In some systems 69 kV may be a transmission voltage while in other systems it is classified as distribution, depending on function.**

# CLASSIFICATION BY OPERATING VOLTAGE

Overhead power transmission lines are classified in the electrical power industry by the range of voltages:

○ **Low voltage** – less than 1000 volts, used for connection between a residential or small commercial customer and the utility.

○ **Medium Voltage** (Distribution) – between 1000 volts (1 kV) and to about 33 kV, used for distribution in urban and rural areas.

○ **High Voltage subtransmission** less than 100 kV; subtransmission or transmission at voltage such as 115 kV and 138 kV), used for sub-transmission and transmission of bulk quantities of electric power and connection to very large consumers.

○ **Extra High Voltage (transmission)** – over 230 kV, up to about 800 kV, used for long distance, very high power transmission.

○ **Ultra High Voltage** – higher than 800 kV.

# DETERMINATION OF TRANSMISSION SYSTEM VOLTAGES

○ **Determining transmission voltages is a matter that requires careful study. Engineers cost out the systems employing several generally standard voltages (and standard materials and equipment). For example, the annual costs for each of the systems: 69 kV, 138 kV, and 230 kV.**

○ **The lowest overall annual expense will generally be found to happen when the annual operating charges are approximately equal to the annual cost of the electrical losses in the system.**

# DETERMINATION OF TRANSMISSION SYSTEM VOLTAGES

- $K_g$: Overall cost of transmission
- $K_v$: The cost of losses on the transmission line
- $K_a$: Cost of transmission equipment
- Where,

$$K_g = K_v + K_a$$

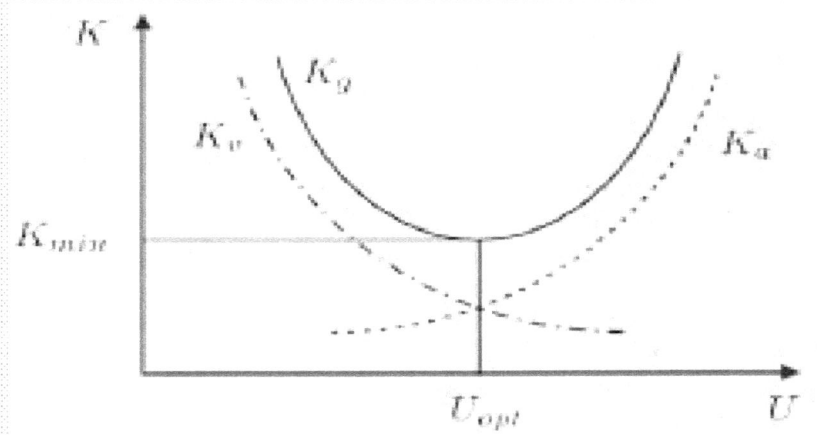

Figure 1 – Transmission System voltage versus cost of transmission

- The economic optimal transmission voltage $U_{opt}$ results at minimal cost $K_{min}$.

# DECIDING BETWEEN OVERHEAD LINES AND UNDERGROUND CABLES

There are two general ways of transmitting electricity:
- Overhead and
- Underground

# COMPARING OVERHEAD AND UNDERGROUND TLs

| | Overhead Lines | Underground Cables |
|---|---|---|
| 1 Conductors | copper or Aluminum conductors, or Aluminum conductors steel reinforced (ACSR) are used | copper or Aluminum conductors, or Aluminum conductors steel reinforced (ACSR) are used |
| 2 Insulation | The insulation is air except at the supports (towers, poles, or other structures), where it may be porcelain, glass, or other material | The conductor is insulated with oil impregnated paper, or a special type of plastic material. |
| 3 Safety | Problems with safety as lines are exposed and prone to natural hazards like rain, wind & lightning | system is safer and less prone to natural hazards like rain, wind and lightning. |
| 4 Initial cost | Transmission systems are costly | Underground systems are much more expensive than overhead systems |
| 5 Flexibility | Overhead system is more flexible than underground system | Underground system have no flexibility |
| 6 Maintenance cost | Maintenance cost is high | Maintenance cost of underground system is very low |
| 7 Frequency of failures | Lines are prone to damage and accidents | Cables are less prone to damage and failure |

# COMPARING OVERHEAD AND UNDERGROUND TLs

| | Overhead Lines | Underground Cables |
|---|---|---|
| 8 Environment | Overhead lines are not considered environmentally friendly | Underground cables are environmentally friendly |
| 9 Fault location & repair | Faults can be located easily and the repair is easy. | if a fault occurs it is very difficult to locate that fault and its repair is difficult and expensive |
| 10 Possible damage | Prone to damage due to thunderstorm and falling objects across the wires | Underground systems are free from interruption of service due to thunder storm lightning and objects falling across the wires. |
| 11 charging current | Small capacitance and low charging current | On account of less spacing between the conductors cables have higher capacitance and draw higher charging current. |
| 13 Joining | Joining is easy | Joining of underground cables is difficult so tapping for loads and service mains is not conveniently possible in underground system. |
| 14 Surge Effects | Overhead lines are prone to severe transients because of exposure and low capacitance | In underground system surge effect is smoothened down as surge energy is absorbed by the sheath |
| 15 Voltage drop | Voltage drop is high because of high inductance | Voltage drop is low because of low inductance |

# TRANSMISSION LINE DESIGN

○ It is desirable, when transmitting large amounts of electric power, to use higher voltages, thereby employing thinner, less expensive conductors that are easier to handle. Low voltages require heavy conductors that are costly and bulky and expensive to install.

○ There is a limit, however, to how high the voltage and how thin the conductors can be. In overhead construction there is the problem of supports—poles, structures, towers.

○ If the conductor is made too thin, it will not be able to support itself mechanically and the cost of additional supports and insulators becomes inordinately high.

31

# TRANSMISSION LINE DESIGN

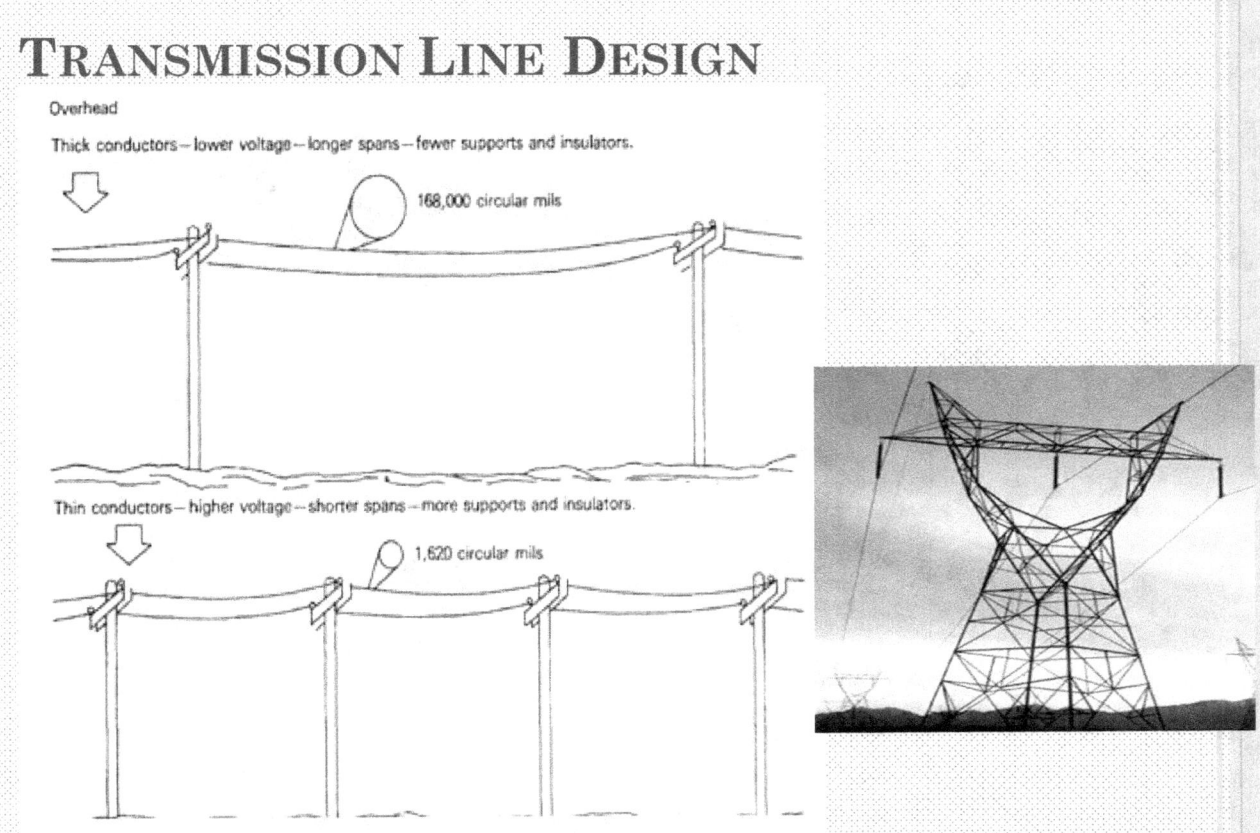

Overhead

Thick conductors—lower voltage—longer spans—fewer supports and insulators.

168,000 circular mils

Thin conductors—higher voltage—shorter spans—more supports and insulators.

1,620 circular mils

Figure 1.2 Practical economics affect the size of a transmission line.

32

# UNDERGROUND CABLES

- Underground construction faces the same economic limitations, and in this case, the expense of insulation.

- A cable must be thoroughly insulated and sheathed from corrosion. The greater the overall size of the cable, the more sheathing becomes necessary and more difficulty experienced in its handling.

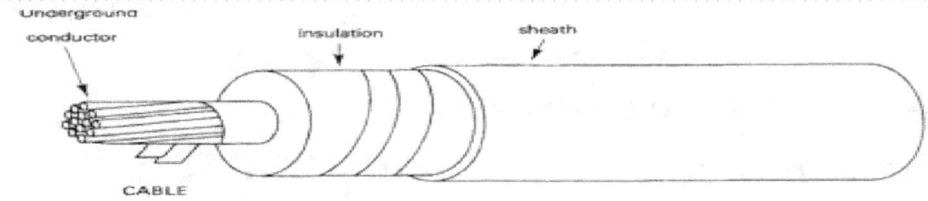

Figure 2 – Design considerations of transmission systems

# UNDERGROUND CABLES

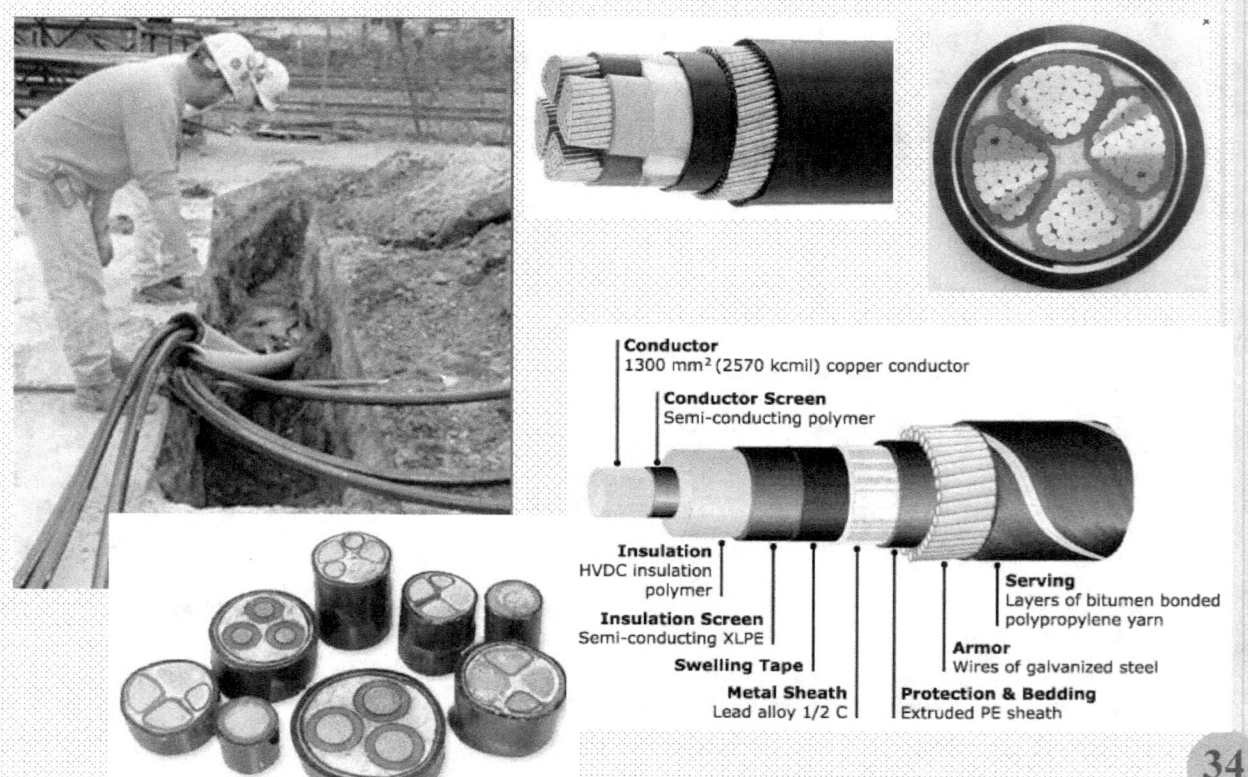

**Conductor**
1300 mm² (2570 kcmil) copper conductor

**Conductor Screen**
Semi-conducting polymer

**Insulation**
HVDC insulation polymer

**Insulation Screen**
Semi-conducting XLPE

**Swelling Tape**

**Metal Sheath**
Lead alloy 1/2 C

**Serving**
Layers of bitumen bonded polypropylene yarn

**Armor**
Wires of galvanized steel

**Protection & Bedding**
Extruded PE sheath

# COST COMPARISON BETWEEN OVERHEAD LINES AND UNDERGROUND CABLES

- The initial cost of underground system is much more expensive than overhead lines.
- For a particular amount of power to be transmitted at a given voltage the underground system costs more than overhead system.
- The approximate cost ratios are as foloows:

| System voltage , KV | 0.4 | 11 | 33 | 66 | 132 | 220 | 400 |
|---|---|---|---|---|---|---|---|
| Cost ratio cable/overhead line | 2 | 3 | 5 | 7 | 9 | 13 | 24 |

# OVERHEAD TRANSMISSION LINES TYPES AND DESIGNS CONSIDERATIONS

- Transmission system design involves the selection of the necessary lines and equipment which will deliver the required power and quality of service for the lowest overall average cost over the service life.
- The system must also be capable of expansion with minimum changes to existing facilities.

# CONDUCTOR SELECTION CONSIDERATIONS

i- General Properties of Transmission Conductors:

Line conductors are one of the main parts of overhead lines. The important characteristics, which the line conductors must have are:

## 1. High tensile strength:

The material for the conductor of an overhead line should have a high tensile strength (high breaking load) so that the spans between transmission line towers can be as long as possible and the sag as small as possible thus reducing the number and height of towers, and number of insulators.

37

## 2. Low resistivity:

The conductor should have low resistivity to reduce the power losses and voltage drop.

## 3. Low cost:

The cost of its installation and maintenance should be low and it should have a long life.

## 4. Low Corrosion:

Conductors must be stable against corrosion.

## 5. Low Skin Effect and Corona Losses:

The structure of the conductor should be such that to minimize the additional losses due to skin effect or corona effect in case of high and extra high voltages.

38

# CONDUCTOR MATERIALS

○ *The final choice of material is often a compromise.*

○ Copper, aluminum, steel and steel-cored aluminum conductors are generally employed in an overhead lines to transmit electrical energy.

○ The following is a list of properties of each of these materials.

## a) Copper.

The most common conductor material used for transmission is hard drawn copper, because it is twice as strong as soft drawn copper. The merits of this metal as a line conductor are:

1. It has a best conductivity in comparison to other metals.

2. It has higher current density, so for the given current density (rating), less cross sectional area of conductor is required and hence it provides lesser cross sectional area to wind loads.

3. The metal is quite homogenous.

4. It has low specific resistance.

5. It is durable and has a higher scrap value

## b) Aluminum:

Next to copper, aluminum is the conductor used in order of preference as far as the conductivity is concerned. Its merits and demerits are:

1. It is cheaper than copper.

2. It is lighter in weight.

3. It is second in conductivity among the metals used for transmission.

4. For same Ohmic resistance, its diameter is about 1.27 times that of copper.

5. Since the diameter of the conductor is more, so it is subjected to greater wind pressure due to which greater is the swing of the conductor and greater is the sag.

6. Since the conductors are liable to swing, so it requires larger cross arms.

7. As the melting point of the conductor is low, so the short circuits and similar effects will damage it.

8. Joining of aluminum is much more difficult than that of any other material.

## c) Steel:

No doubt it has got the greater tensile strength, but it is least used for transmission of electrical energy as it has got high resistance. Bare steel conductors are not used since they deteriorate rapidly owing to rusting. Generally galvanized steel wires are used where high strength is desired. It has the following properties:

1. It has high internal reactance.

2. It is lowest in conductivity,

3. It is much subjected to eddy current and hysteresis losses.

4. In a damp atmosphere it rusts quickly.

Hence its use is limited and the main application is

## d) Aluminum conductor with steel reinforced (ACSR).

An aluminum conductor having a central core of galvanized steel wires is used in stranded conductors for high voltage transmission purposes. This is done to increase the tensile strength of aluminum conductors. The galvanized steel core is covered by one or more strands of aluminum wires. The steel conductors used are galvanized in order to prevent rusting and electrolytic corrosion.

Examples of such conductors are shown in Figure 3.

# ACSR Lines

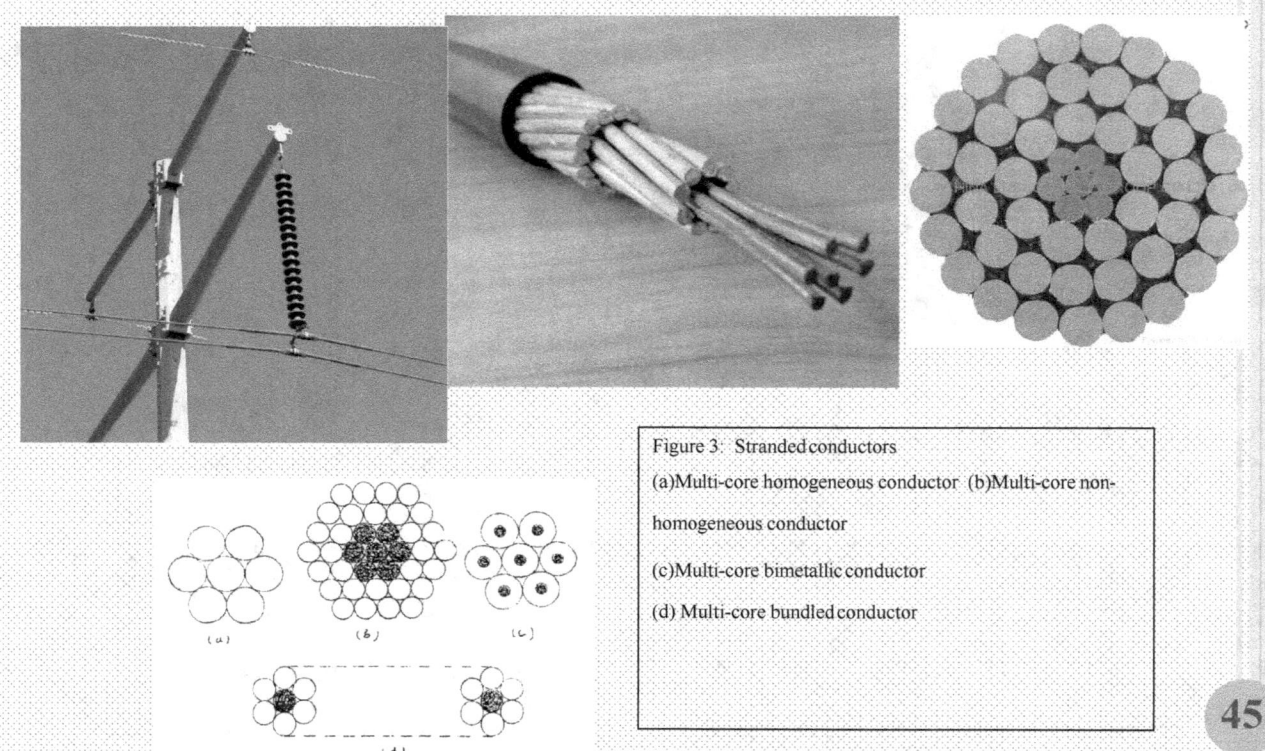

Figure 3: Stranded conductors
(a)Multi-core homogeneous conductor  (b)Multi-core non-homogeneous conductor
(c)Multi-core bimetallic conductor
(d) Multi-core bundled conductor

# INSULATORS

- **Insulators must support the conductors and withstand both the normal operating voltage and surges due to switching and lightning. Insulators are broadly classified as either**
- **Insulators are usually made of wet-process *porcelain* or toughened glass , with increasing use of glass-reinforced polymer insulators.**

1) **pin-type, which support the conductor above the structure**

# Pin Type Insulator

As the name suggests, the pin type insulator is mounted on a pin on the cross-arm on the pole. There is a groove on the upper end of the insulator. The conductor passes through this groove and is tied to the insulator with annealed wire of the same material as the conductor.

Pin type insulators are used for transmission and distribution of communications, and electric power at voltages up to 33 KV. Insulators made for operating voltages between 33 KV and 69 KV tend to be

2) suspension type, where the conductor hangs below the structure. The invention of the strain-insulator was a critical factor in allowing higher voltages to be used.

Suspension insulators are made of multiple units, with the number of unit insulator disks increasing at higher voltages. The number of disks is chosen based on line voltage, lightning withstand requirement, altitude, and environmental factors such as fog, pollution, or salt spray.

# Suspension Insulator

For voltages greater than 33 KV, it is a usual practice to use suspension type insulators, consisting of a number of glass or porcelain discs connected in series by metal links in the form of a string. The conductor is suspended at the bottom end of this string while the top end is secured to the cross-arm of the tower. The number of disc units used depends on the voltage.

## LINE SUPPORTS

Line supports can be made up of poles or towers and has to perform the following functions:

a. Keep appropriate spacing between the conductors.

b. Maintain the specified ground clearance.

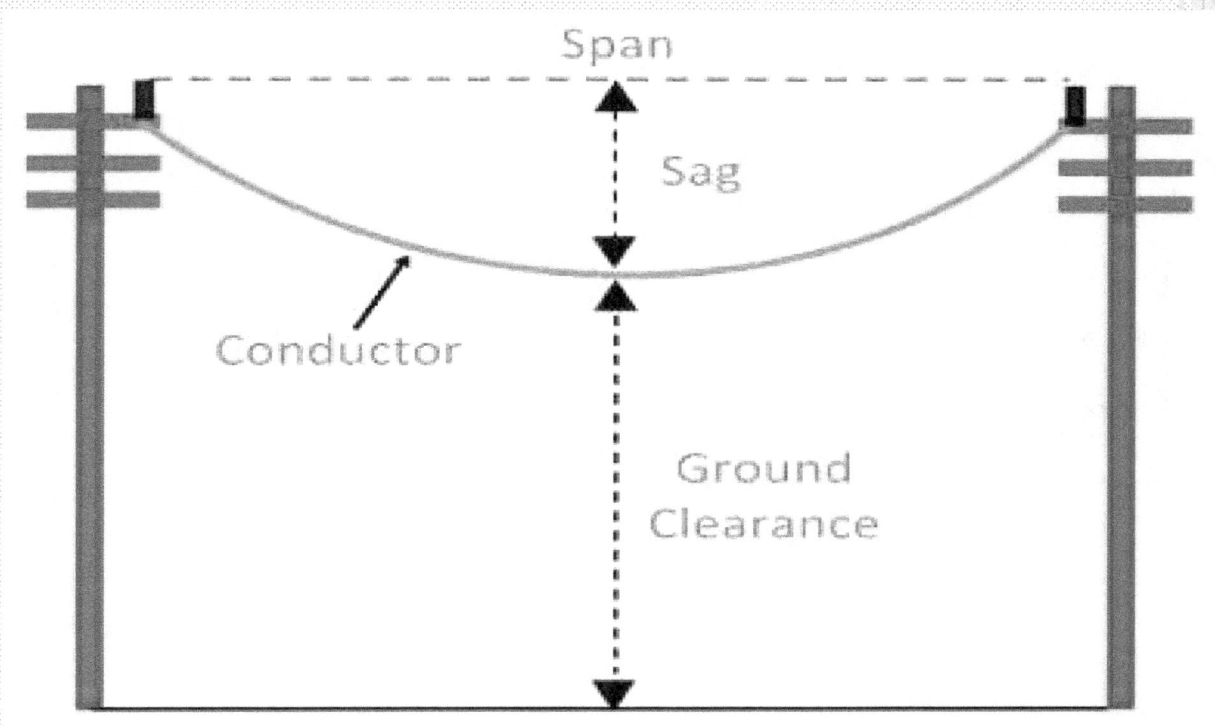

## LINE SUPPORTS

### *Poles:*

Poles are usually used for short spans in low voltage lines. Various types of poles are used, such as: Wooden poles, concrete poles and steel poles.

_Towers:_

_High voltage lines require large air and ground clearances. Steel towers were developed for such lines where very long spans are essential. They are classified as:_

_a) Self - supporting towers._

_b) Guyed towers._

| Monopole | Self-Supporting | Guyed |
| 100—200 feet tall | 100—400 feet tall | 100—2,150 feet tall |

## SELF-SUPPORTING TOWERS

These can be:

○ **Wide-base:** In wide base towers, _lattice_ type construction with bolted connection is adopted. Each leg has a separate foundation.

○ Narrow-base towers. These use latticed construction of angle, channel or tubular steel sections with bolted or welded connections. They use less steel in comparison with wide-base towers, but use more costly foundation.

## SELF-SUPPORTING TOWERS

○ A narrow-base tower requires lesser steel in comparison with a wide-base tower, but its cost of foundation is more. The selection between the two has to be made on the basis of comparison between the cost of material and foundation.

○ Figure 4 shows various types of steel towers.

51

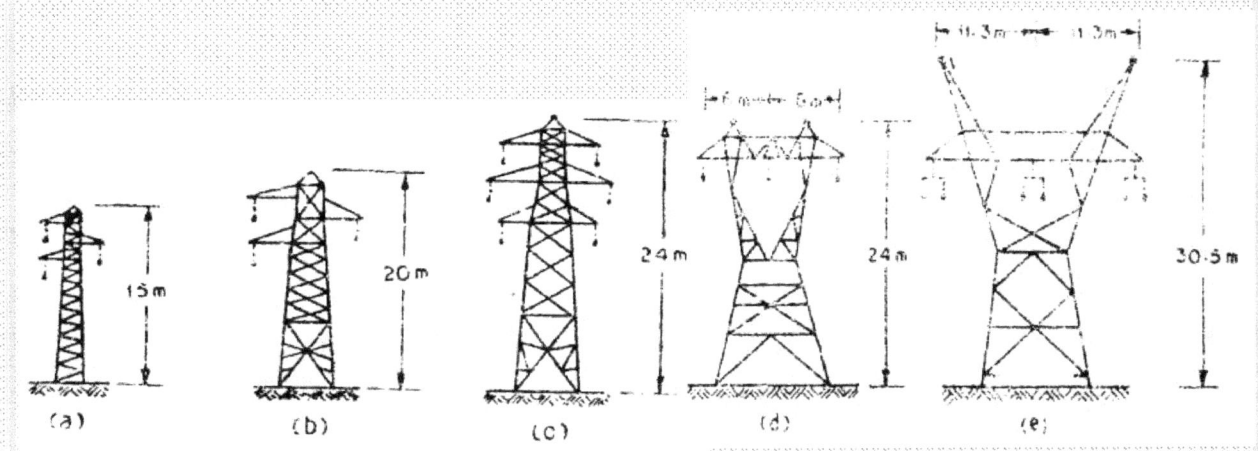

Figure 4 - Steel Towers:

(a) 33 kV, narrow base Single circuit

(b) 66 kV, broad base, single circuit

(c) 132 kV, double circuit, broad base

(d) 220 kV Cat's head, single circuit

(e) 400 kV, single circuit with two sub-conductors per phase

52

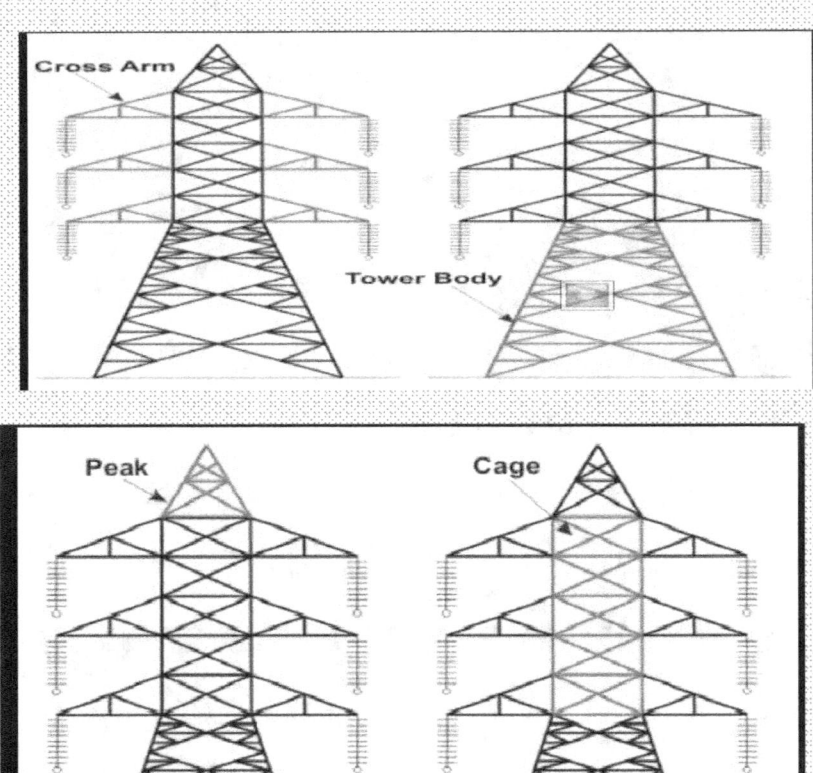

# WAIST-TYPE TOWER

**Waist-type tower**

This is the most common type of transmission tower. It's used for voltages ranging from 110 to 735 kV. Because they're easily assembled, these towers are suitable for power lines that cross very uneven terrain.

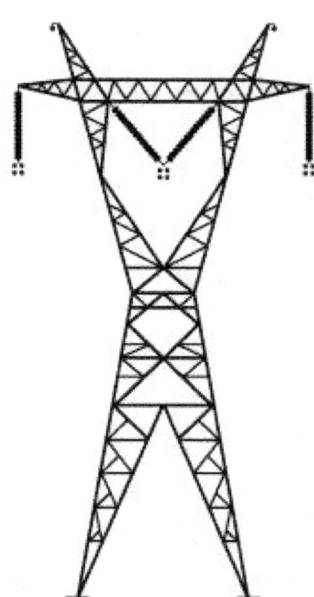

## Double-circuit tower

This small-footprint tower is used for voltages ranging from 110 to 315 kV. Its height ranges from 25 to 60 metres.

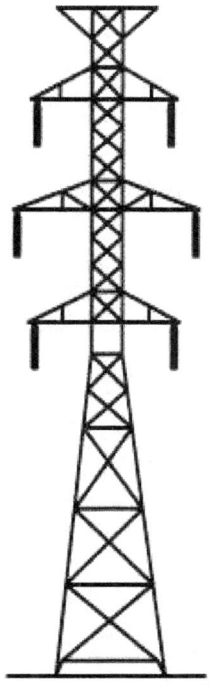

## Guyed-V tower

This tower is designed for voltages ranging from 230 to 735 kV. It's used mainly for power lines leaving the La Grande and Manic-Outardes hydroelectric complexes. The guyed-V tower is more economical than the double-circuit and waist-type towers.

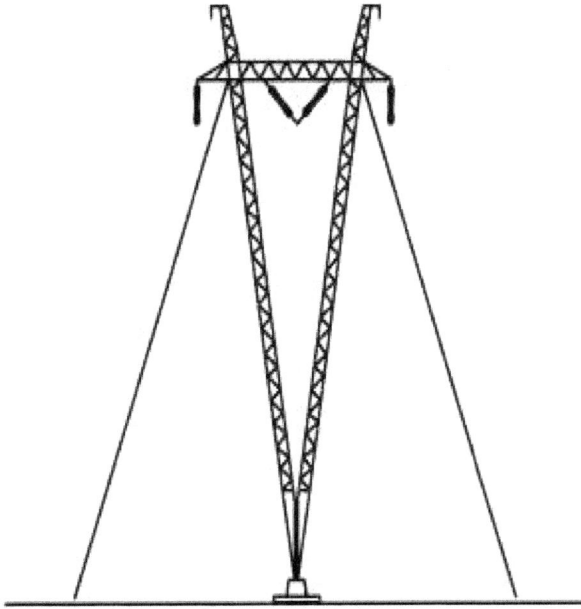

## Tublar steel pole

Featuring a streamlined, aesthetic shape, this structure is less massive than other towers, allowing it to blend easily into the environment. For this reason, it's being used more and more in urban centres. Measuring between 27 and 45 metres in height, it's suitable for voltages ranging from 110 to 315 kV.

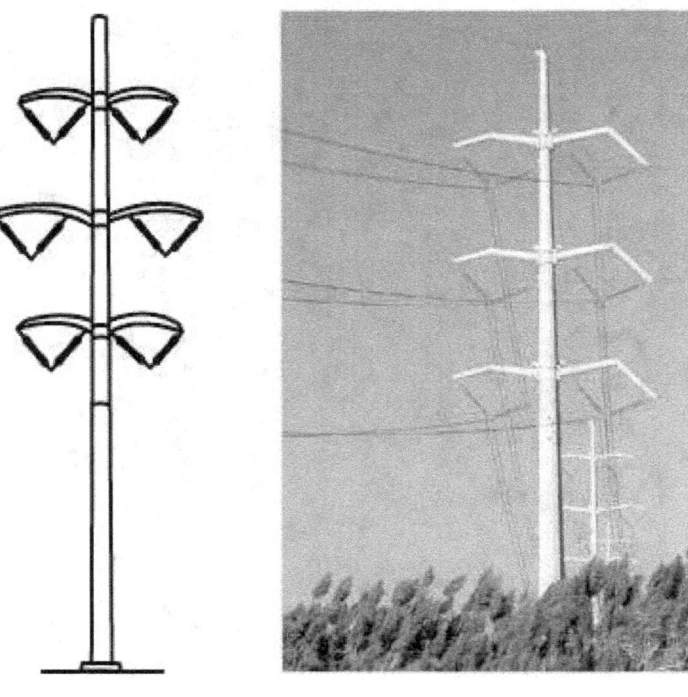

## Guyed cross-rope suspension tower

With its simple design, this tower is easy to assemble. It's used on some sections of power lines leaving the La Grande complex and supports 735-kV conductors. This type of structure requires less galvanized steel than the guyed-V tower, making it lighter and less costly.

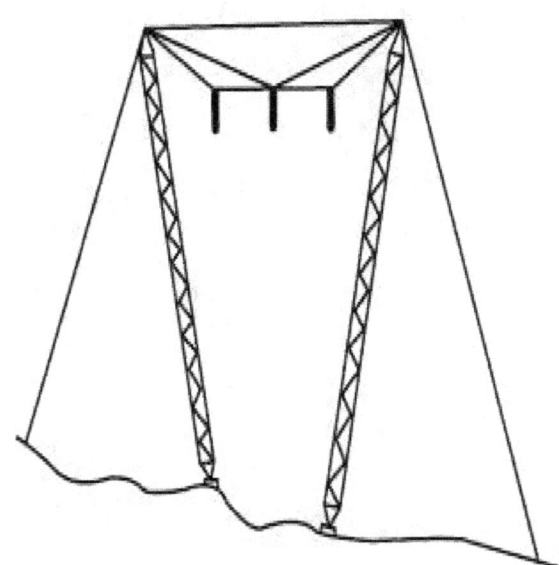

# EEL 2023

# Power Generation and Transmission

## Chapter 5

### Transmission System Stability

# Synchronous Generator

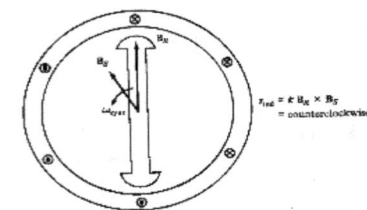

**Principle of Operation**
1)The DC current applied to rotor winding produces magnetic field $B_R$
2) The rotor turned by primemover so there will be rotating magnetic field within the machine.
3) The rotating magnetic field induces 3 phase set of voltages within the stator windings.

## Equivalent Circuit

The relation between the speed of rotation and the frequency of the synchronous machine

$$f_e = \frac{n_m P}{120}$$

where  $f_e$ = electrical frequency, Hz
  $n_m$ = mechanical speed of magnetic field, r/min (= speed of rotor for synchronous machines)
  $P$ = number of poles

# The output voltage

$$E_A = K\phi\omega$$

## The equivalent circuit

$$V_\phi = E_A - jX_S I_A - R_A I_A$$

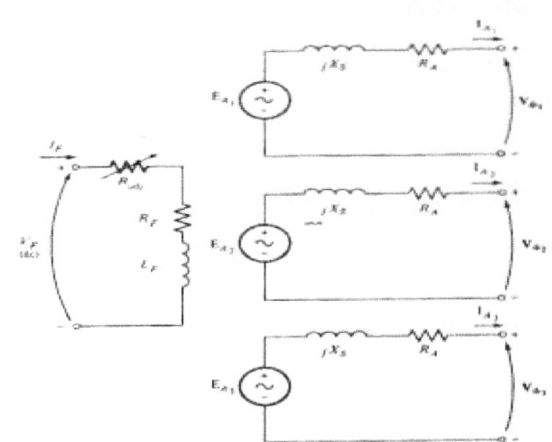

# The windings can be connected in Star or Delta

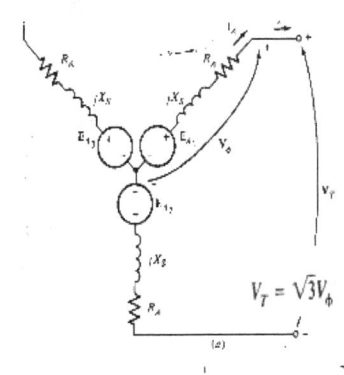

$$V_T = \sqrt{3}V_\phi$$

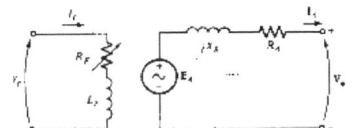

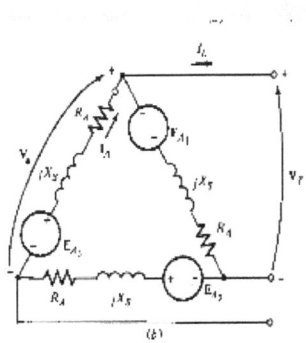

$$P_{elec} = 3V_\varphi I_a \cos(\varphi)$$

$$Q_{3\varphi} = 3V_\varphi I_a \sin(\varphi)$$

# Power System Stability

- Power system stability is directly related to synchronous generator stability.

- A synchronous machine is said to be stable if, under steady state conditions, it is operating in equilibrium.

- Equilibrium denotes perfect power balance between input and output powers.

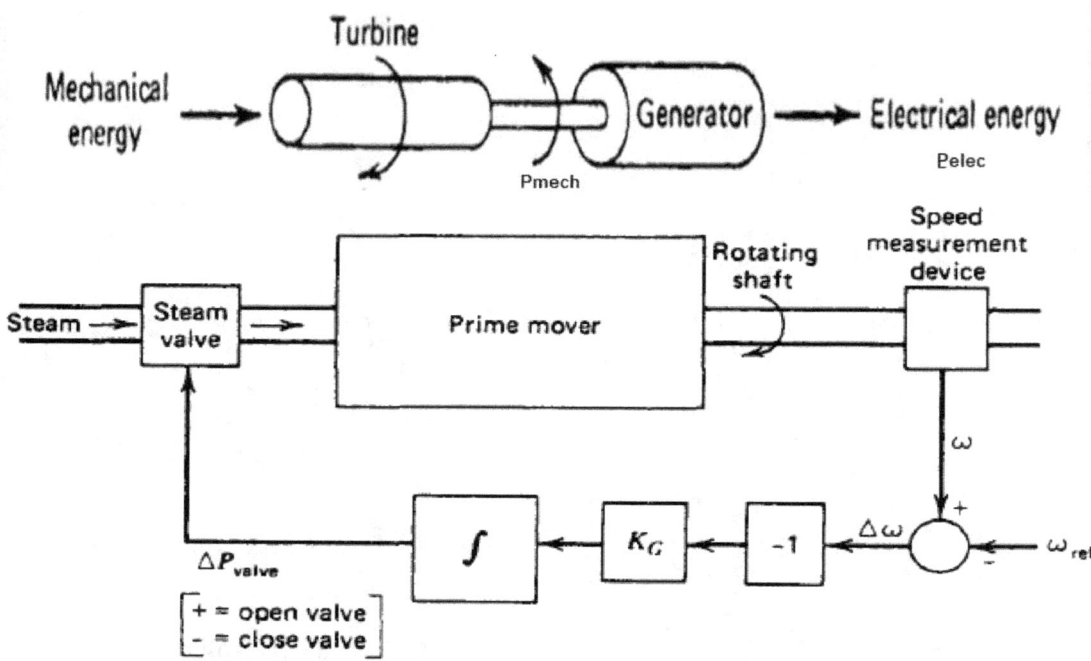

# Swing Equation

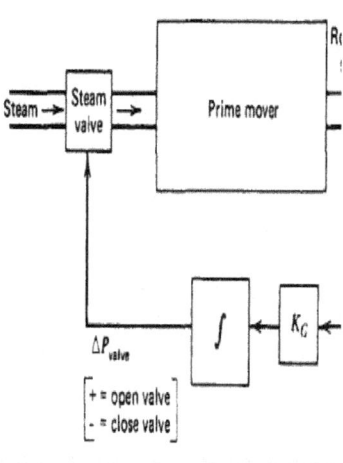

*When No Damping*

$$P_{mech} - P_{elec} = M \frac{d\omega}{dt}$$

$$\omega = \frac{d\delta}{dt}$$

*When Damping*

$$P_{mech} - P_{elec} = D\omega + M \frac{d\omega}{dt}$$

**M: Angular momentum of machine**

- When electrical load Pe increases, speed will decrease. So governor will increase mechanical power Pm so that Pe=Pm
- When electrical load decrease, speed will increase. So that governor will decrease mechanical power Pm so that Pe=Pm
- But the governor slow, and until governor will act, there can be problem of instability

# Power System Stability

- A synchronous machine connected to an infinite bus is said to be working in a stable condition, if it is in *synchronism*, or in step, with the bus. Unstable operation denotes loss of synchronism or falling out of step.

- *Synchronism* - The tendency of a power system to develop forces to maintain equilibrium (i.e. to stay synchronized).

➢ Stability is a condition of equilibrium between opposing forces

➢ Instability results when a disturbance causes imbalance between the opposing forces.

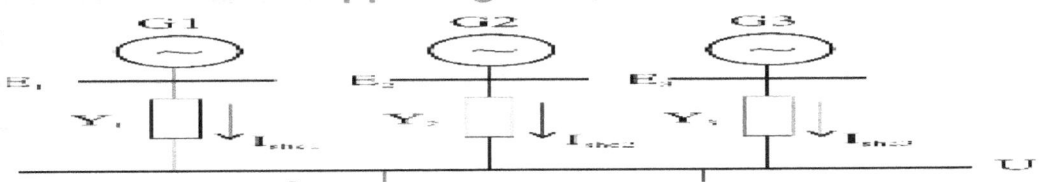

* Swing equation Controlling Stability

$$P_{mech} - P_{elec} = M\frac{d\omega}{dt}$$

M: Angular Momentum of Machine

$P_{mech} = P_{elec}$ in Steady State

So acceleration & deceleration $\frac{d\omega}{dt} = 0$

$$\omega = \frac{d\delta}{dt}$$

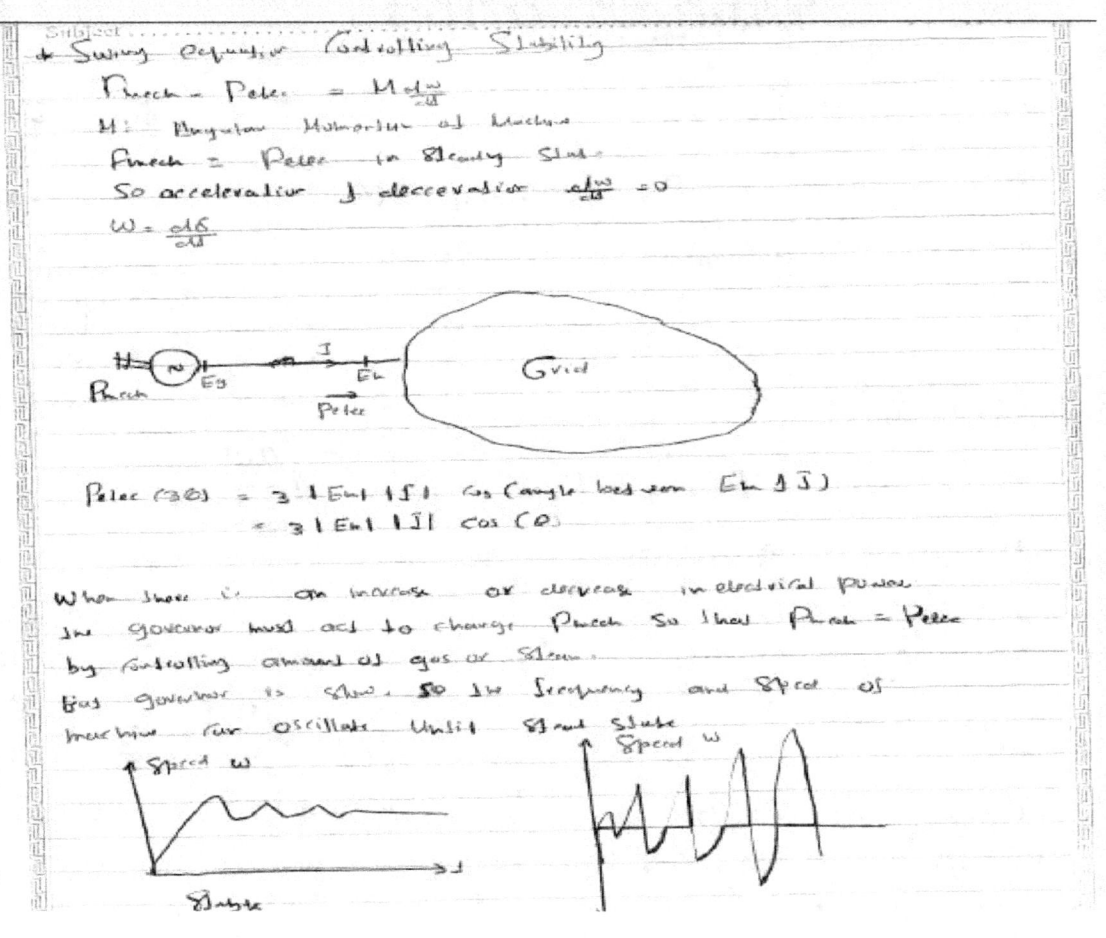

$$P_{elec} (3\emptyset) = 3 |E_m| |I| \cos(\text{angle between } E_m \text{ & } I)$$
$$= 3 |E_m| |I| \cos(\theta)$$

When there is an increase or decrease in electrical power the governor must act to change $P_{mech}$ so that $P_{mech} = P_{elec}$ by controlling amount of gas or steam.

This governor is slow, so the frequency and speed of machine can oscillate until steady state

Per phase equation (errors)

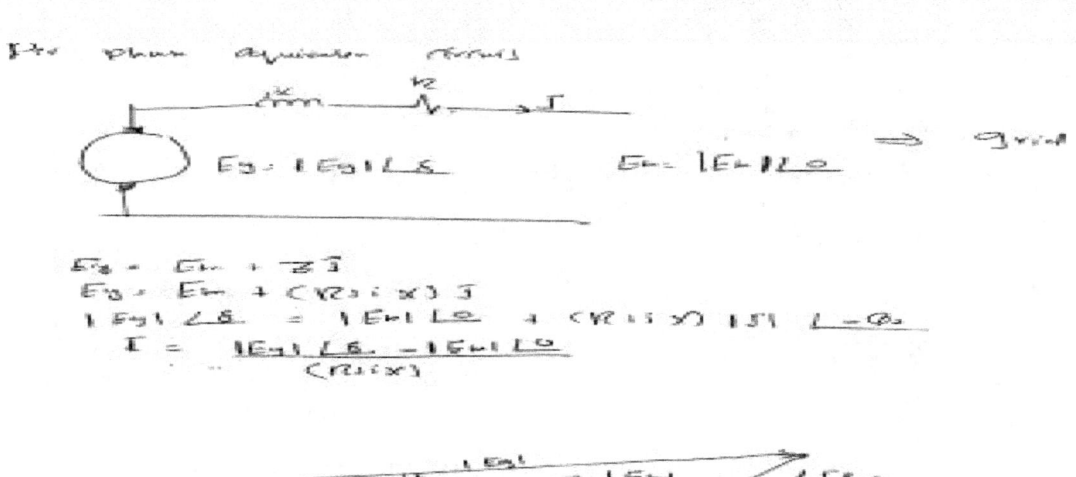

$E_g = E_m + ZI$

$E_g = E_m + (\sqrt{R} + jx) I$

$|E_g| \angle \delta = |E_m| \angle 0 + (\sqrt{R} + jx) |I| \angle -\theta_v$

$I = \dfrac{|E_g| \angle \delta - |E_m| \angle 0}{(\sqrt{R} + jx)}$

$P_{elec} = P_m \cos(\emptyset) = |E_m| |I| \cos\theta_v$

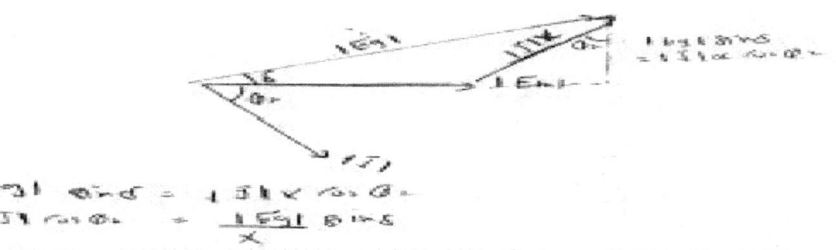

$|E_g| \sin\delta = |I| x \cos\theta_v$

$|I| \cos\theta_v = \dfrac{|E_g| \sin\delta}{x}$

$$P_{elec}(\delta) = P_m(\delta) = \frac{|E_h||E_g|}{X} \sin\delta$$

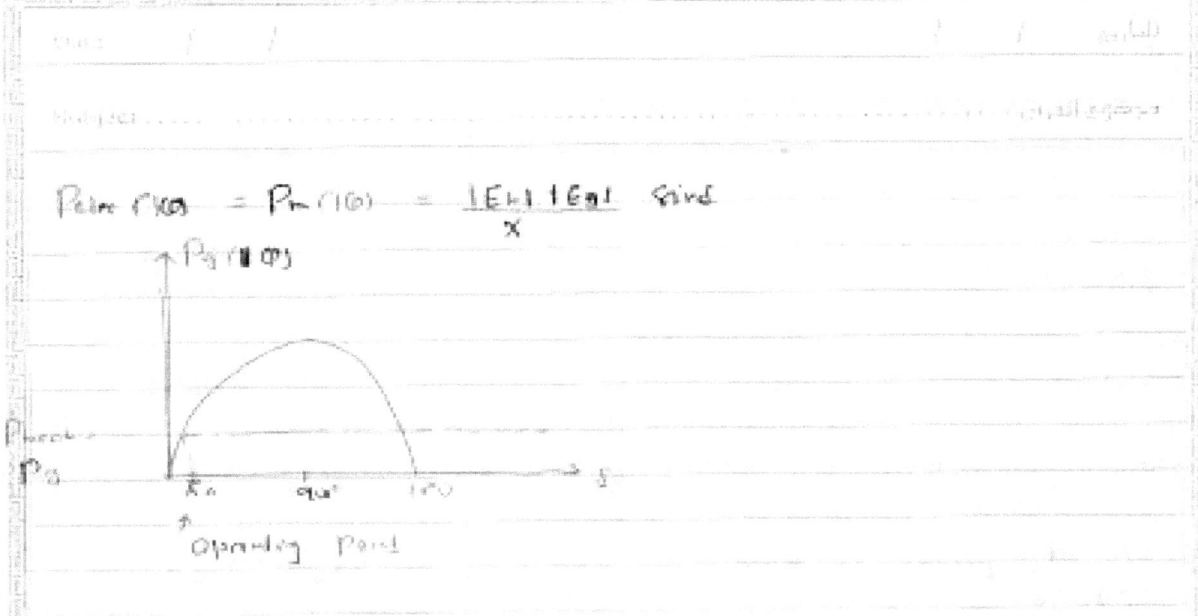

Operating Point

Example

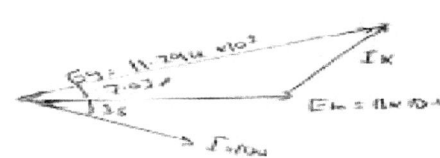

$jX = j5$

$I = 100\angle -35$

$E_h = 11\,kv\,\angle 0$

$E_g =$

$E_g = E_h + jX\bar{I}$

$= 11\times10^3 \angle 0 + j5 \times 100\angle -35$

$= 11\times10^3 \angle 0 + 5\angle 90 \times 100\angle -35$

$= 11\times10^3 + 500\angle 55$

$= 11\times10^3 + 286.7886 + j409.576$

$= 11286.788 + j409.576$

$= 11.294\times10^3 \angle 7.036$

$E_g = 11.294\times10^3$, $7.036$, $35$, $I_x$, $E_h = 11\times10^3$, $I\cdot100$

$P_m(\delta) = P_g(\delta) = |E_h||I|\cos(\text{angle between } E_h \text{ \& } I)$

$= 11\times10^3 \times 100 \cos(35)$

$= 0.9\times10^6 = 0.9\,MW$

$P_g(\delta), P_m(\delta)$

$\delta = 7°$

$P_m(\delta) = P_g(\delta) = \frac{|E_g||E_h|}{X}\sin\delta$

$= \frac{11000 \times 11.294\times10^3}{5}\sin(7.036) = 0.9\times10^6\,W$

$P_{g,max}(\delta) = P_{m,max}(\delta) = \frac{11\times10^3 \times 11.294\times10^3}{5} = 24.8468\,MW$

# Power System Stability

- A Power system is stable when it can regain a state of operating equilibrium after being subjected to a physical disturbance.

### A Simple Mechanical Analogy of Stability

*Stable Equilibrium:*

- A marble rests at the bottom of the bowl. Displace the marble from this resting position, it it will move up and down and settles again at the bottom.
- This is a stable equilibrium.
- Gravity is the restoring force in this case.

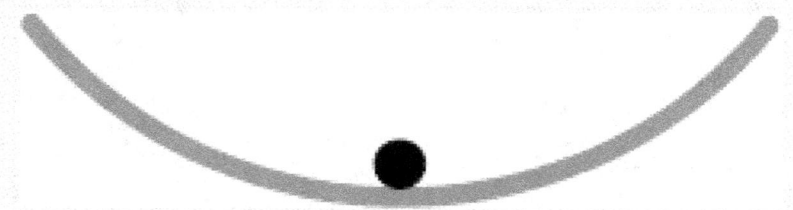

## A Simple Mechanical Analogy of Stability:

### *Unstable Equilibrium:*

- In this case, the marble is balanced at the top of the bow. Any movement will cause loss of equilibrium.

- This type of equilibrium is unstable.

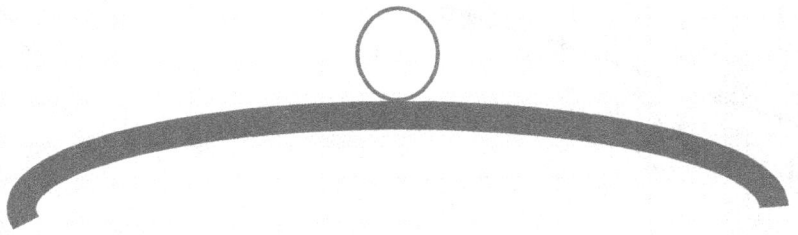

## Steady-State Stability

- **To study the steady-state stability, consider the two-machine system represented by the simple two-terminal circuit shown in the following figure.**

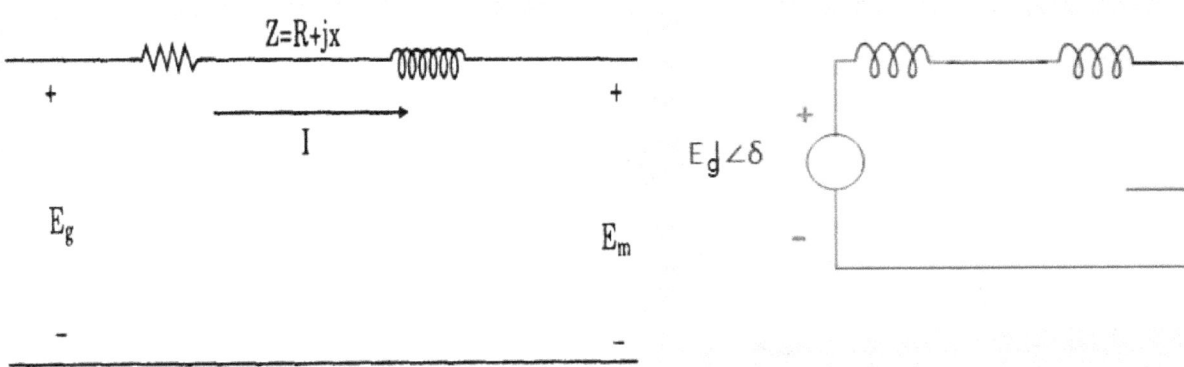

- **The phasor diagram of the circuit is shown in the following figure:**

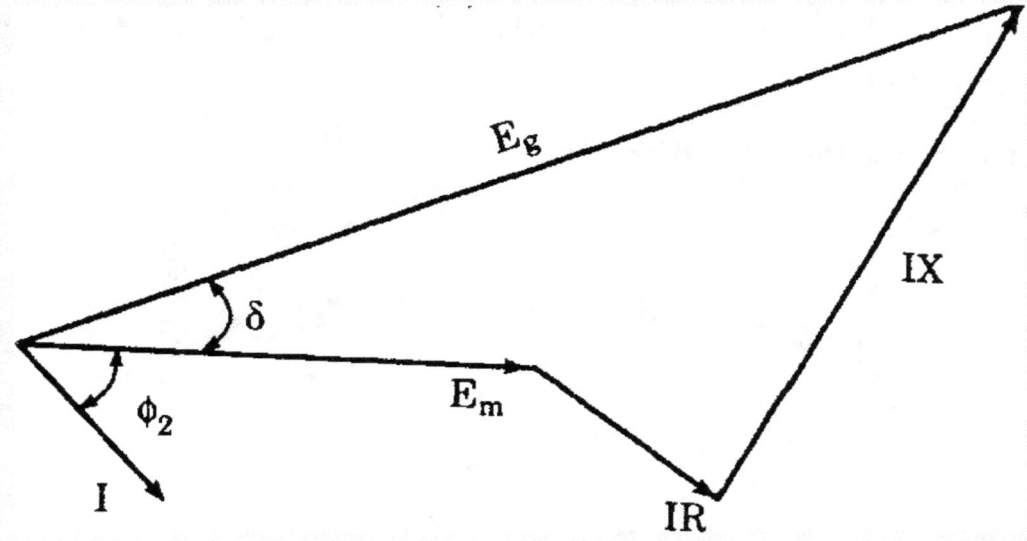

The power delivered to the load, $P_m$, is:

$$P_m = E_m I \cos\Phi_2 \qquad (1)$$

where $\Phi_2$ is the phase angle between $E_m$ and $I$.

The current $I$ is given by:

$$I = \frac{E_g - E_m}{Z} \qquad (2)$$

**The current, *I*, can also be written as:**

$$I = \frac{E_g \angle \delta - E_m \angle 0°}{|Z| \angle \phi_z} = \frac{E_g}{|Z|} \angle \delta - \phi_z - \frac{E_m}{|Z|} \angle -\phi_z \qquad (3)$$

**where**

$E_g$ = sending end voltage

$E_m$ = receiving end voltage

$\delta$ = phase angle between $E_g$ and $E_m$

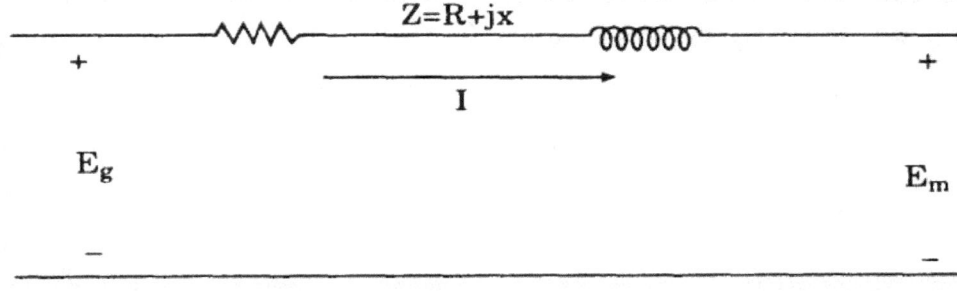

***I Cos$\Phi_2$* can be written as:**

$$I \cos \phi_2 = \frac{E_g}{|Z|} \cos (\delta - \phi_z) - \frac{E_m}{|Z|} \cos (-\phi_z) \qquad (4)$$

**Substitute equation (4) into equation (1) we get:**

$$P_m = \frac{E_g E_m}{|Z|} \cos (\delta - \phi_z) - \frac{E_m^2 R}{|Z|^2} \qquad (5)$$

# Power Angle δ

- **If the line resistance R is negligible, it can be shown that P can be written as:**

$$P_g = P_m = \frac{E_g\,E_m}{X}\sin\delta \qquad (6)$$

This is the _**power angle equation**_ of the system. where

   $E_g$ = sending end voltage

   $E_m$ = receiving end voltage

   δ  = phase angle between $E_g$ and $E_m$

   $P_g$ is the power produced by the generator  $P_m$ is the power delivered to the load.

# Power Angle curve

The real power variation with power angle for     both generator and grid for constant values of Eg, Em and X are shown below.

This curve is known as _**power angle curve**_.

If R = 0, the maximum power is given by:

$$P_{g,\,max} = P_{m,\,max} = \frac{E_g\,E_m}{X}$$

This maximum power occurs at 90   (7)

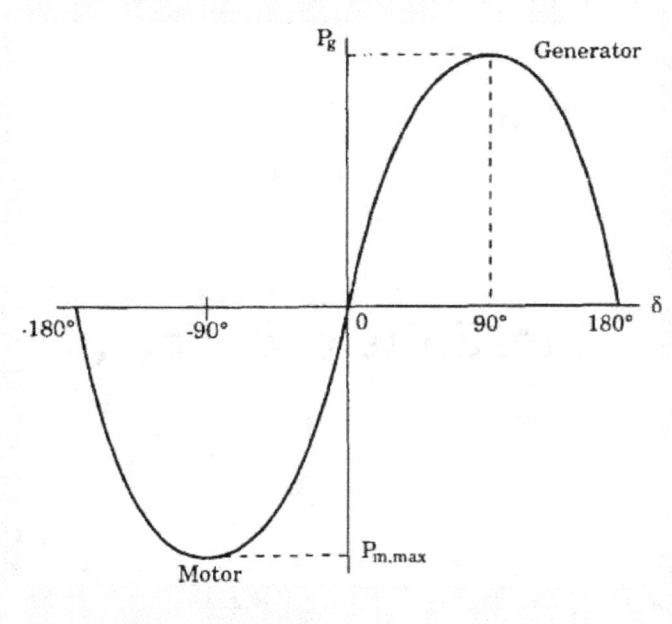

# Power Angle (δ)

- **In a power system, the parameter that determines the stability of the system is the power angle (δ ) between generators or systems.**

*When No Damping*

$$P_{mech} - P_{elec} = M\frac{d\omega}{dt}$$

$$\omega = \frac{d\delta}{dt}$$

*When Damping*

$$P_{mech} - P_{elec} = D\omega + M\frac{d\omega}{dt}$$

# Power System Disturbances

## _Small disturbances:_

- These can be due to load changes that occur continually. The system adjusts to such changing conditions without loss of stability.

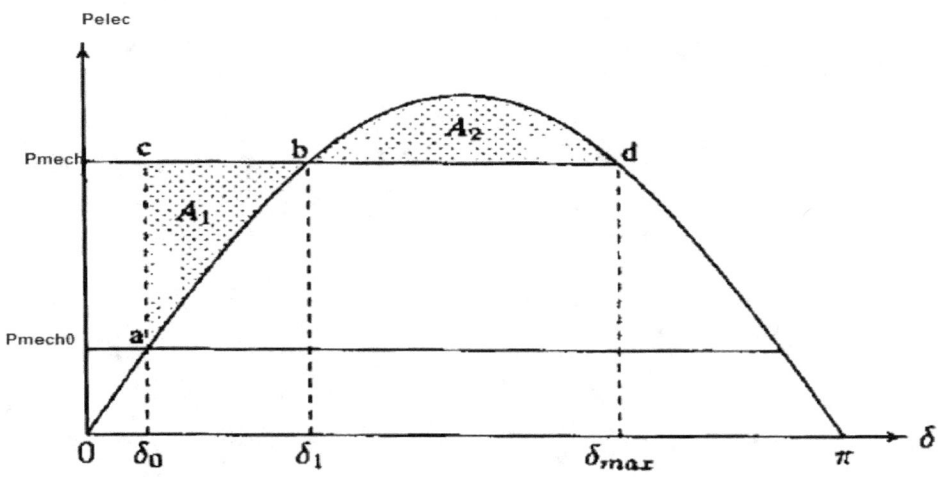

## _Large Disturbances:_

- Large disturbances, such as a short-circuit on a transmission line or loss of a large generator can occur. A stable system should survive such disturbances.

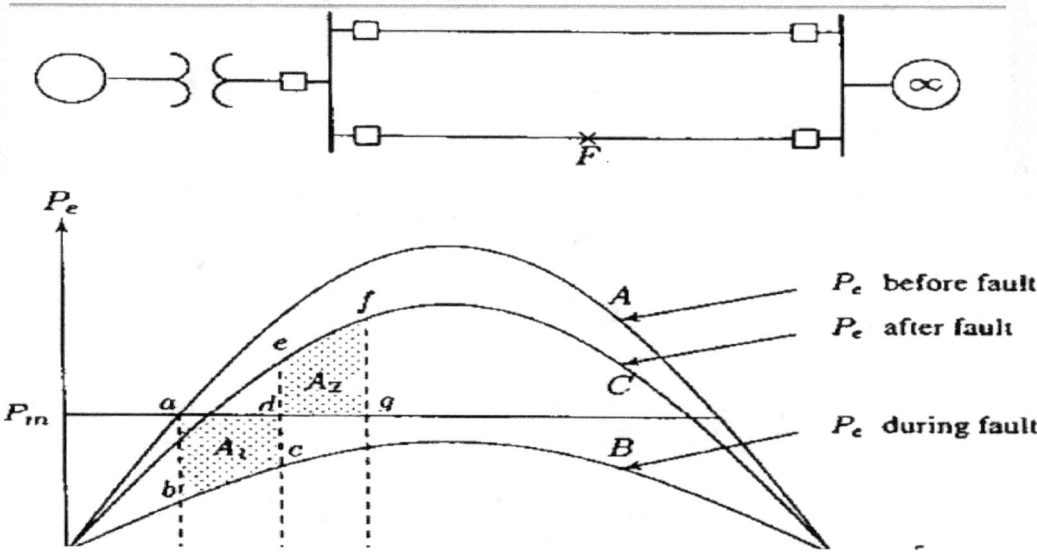

# Rotor Angle Stability

- *Rotor angle stability* means that interconnected synchronous machines should remain in synchronism under normal and abnormal operating conditions.

- *Instability* occurs when the power angle of some generators increases leading to their loss of synchronism.

## Maintaining Stability

- Q. *How do interconnected system maintain stability ?*

- When synchronous machines in a power system start to accelerate or decelerate, due to disturbances, restoring forces are developed which will act so as to keep the speed constant.

# Methods for increasing stability limits

If R = 0, the maximum power is given by:

$$P_{g,\,\text{max}} = P_{m,\,\text{max}} = \frac{E_g\, E_m}{X}$$

- An increase in generator excitation increases the maximum power that can be transferred between the machines, and stability.

- If the internal voltages of generators are increased without an increase in the power transferred, the torque angle decreases.

- Any reduction in the reactance of the network increases the stability limit.

# Methods for increasing stability limits

- Stability of a power system can be increased by using two parallel transmission lines instead of one.

- Series capacitors can be used on transmission lines to decrease the line reactance and raise the system stability limit.

# Maintaining stability limits

- If the system is disturbed, one of two things can happen:
- If the system is stable, it will reach a new equilibrium state (new power angle) with practically the entire system intact
- If the system is unstable, it will result in a progressive increase in the load angle of generators, or a progressive decrease in bus voltages. An unstable system condition could lead to a shut-down of a major portion of the power system.

- Generators in synchronism develop restoring forces that slow down a generator that has sped up and to speed up a generator that has slowed down.
- Such restoring forces exist. The force results from the fact that a generator whose power angle is ahead of others must supply additional power (thus tending to restrain the turbine more), whereas the generator whose power angle is behind supplies less power (thus relieving the restraint on the turbine).

- This figure shows the limits of steady-state stability. The slope of the Sine curve is steep for small $\delta_{12}$ and gets flatter as $\delta_{12}$ increases. A flatter slope means that, for a given increment in $\delta_{12}$, there is only a small increment in P.

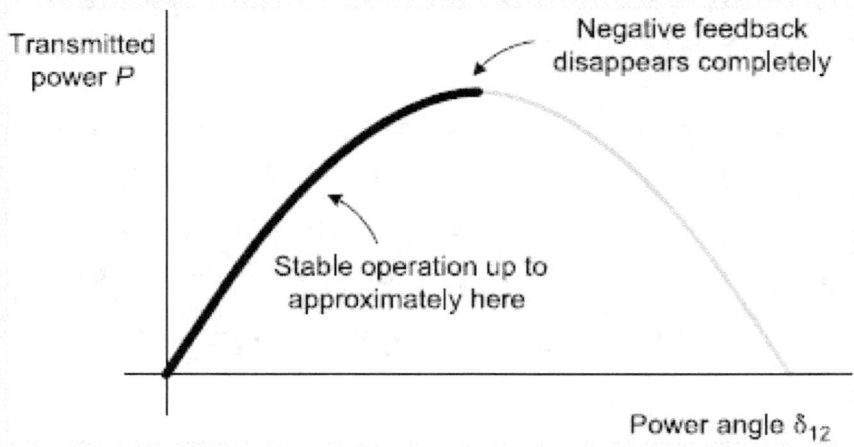

- For a stable generator, increment in P needs to be large. This is analogous to the slope of the sides of the bowl with the marble:

- A deep bowl (a) means strong restoring forces — stability assured.

- ❖ A shallow bowl (b) means weak restoring force - stability not assured.

- **Therefore, it is preferable to operate with a small $\delta_{12}$ where the slope of Sin $\delta_{12}$, and the incremental change in P, is large.**

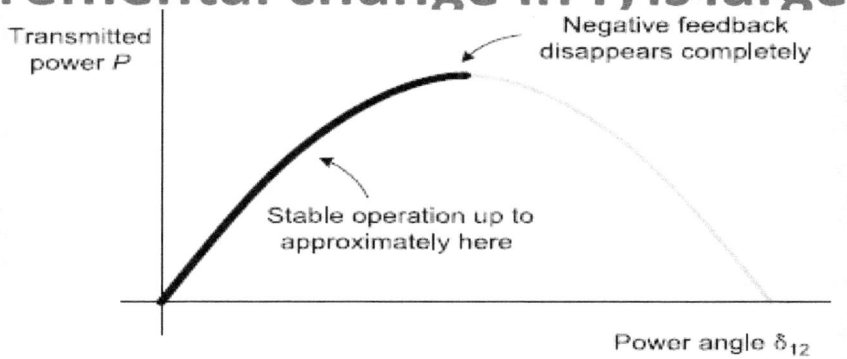

- **Stability limit is when the slope of Sin $\delta_{12}$ is just steep enough. Based on experience, power engineers generally consider 40° to 50° a reasonable limit on $\delta_{12}$**

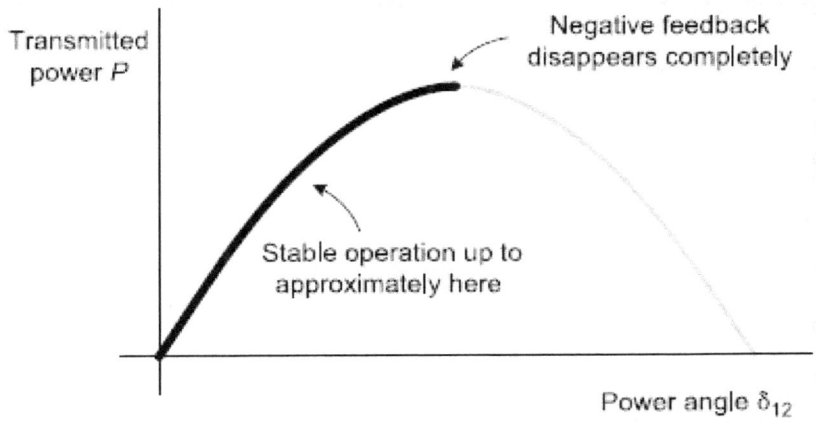

# Steady-State Stability of Interconnected Systems

- Steady-state stability is not only related to interconnected synchronous generators but also to power systems interconnected by transmission lines.

$$P = \frac{V_1 V_2}{X} \sin \delta_{12}$$

Where:

- $\delta_{12}$ is the difference in power angles between the sending and receiving end of the line
- $V_1$ and $V_2$ are the voltage magnitudes at either end of the line,
- X is the reactance of the line in between.

- For short lines, the reactance X is small, so that a small $\delta_{12}$ still results in a large amount of power transmitted.

- Consequently, if $\delta_{12}$ is allowed to take it's maximum value in such a line, the power transmitted could easily exceed the line's thermal capacity.

- For long lines the reactance becomes more significant and a high $\delta_{12}$ may well be reached before the thermal limit of the line.

This figure shows the stability limits and thermal limits of transmission lines as a function of line length.

The label $P_{12}/P_{SIL}$ is the real power transmitted between the two ends of the line, expressed as a ratio of the actual power in watts and the surge impedance loading, which is a characteristic of a given line

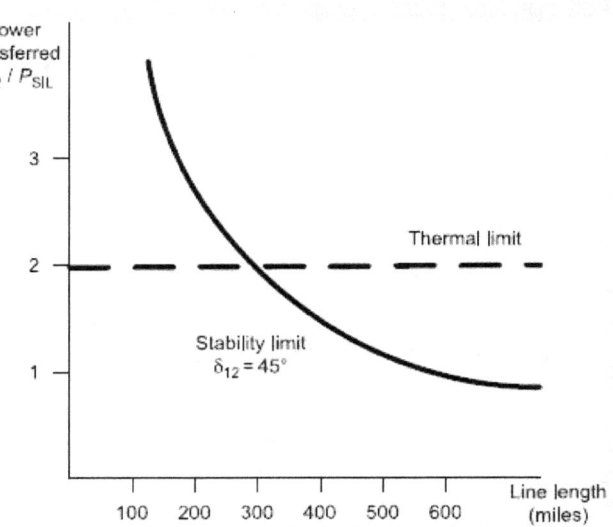

# EEL 2023

# POWER GENERATION AND TRANSMISSION

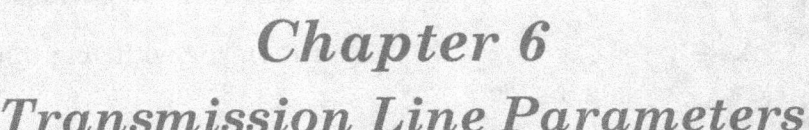

## Chapter 6
### Transmission Line Parameters

# Transmission Lines

## *Introduction*

➢ Generators and loads are connected together through transmission lines transporting electric power from one place to another. Transmission line must, therefore, take power from generators, transmit it to location where it will be used, and then distribute it to individual consumers.

➢ The power capability of a transmission line is proportional to the square of the voltage on the line. Therefore, very high voltage levels are used to transmit power over long distances. Once the power reaches the area where it will be used, it is stepped down to a lower voltages in distribution substations and then delivered to customers through distribution lines.

Distribution line with no ground wire.

Dual 345 kV transmission line (USA Power System)

There two types of transmission lines:
1- overhead lines and
2- buried cables.

➤ An overhead transmission line usually consists of three conductors or bundles of conductors containing the three phases of the power system. The conductors are usually aluminum cable steel reinforced (ACSR), which are steel core (for strength) and aluminum wires (having low resistance) wrapped around the core.

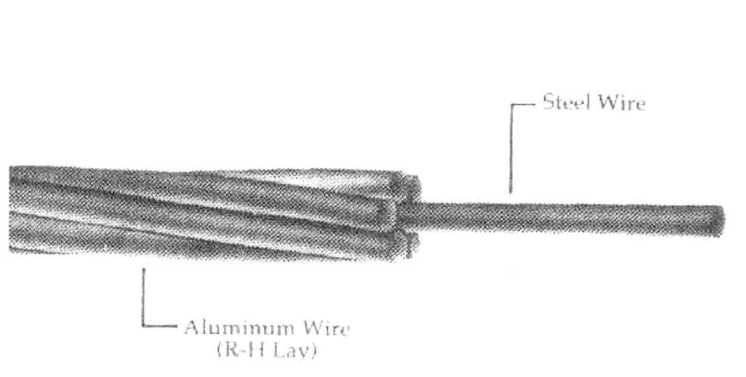

Steel Wire

Aluminum Wire
(R-H Lay)

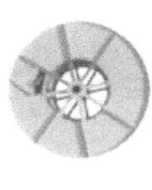

➢ In overhead transmission lines, the conductors are suspended from a pole or a tower via insulators.

➢ In addition to phase conductors, a transmission line usually includes one or two steel wires called ground (shield) wires. These wires are electrically connected to the tower and to the ground, and, therefore, are at ground potential.

➢ In large transmission lines, these wires are located above the phase conductors, shielding them from lightning.

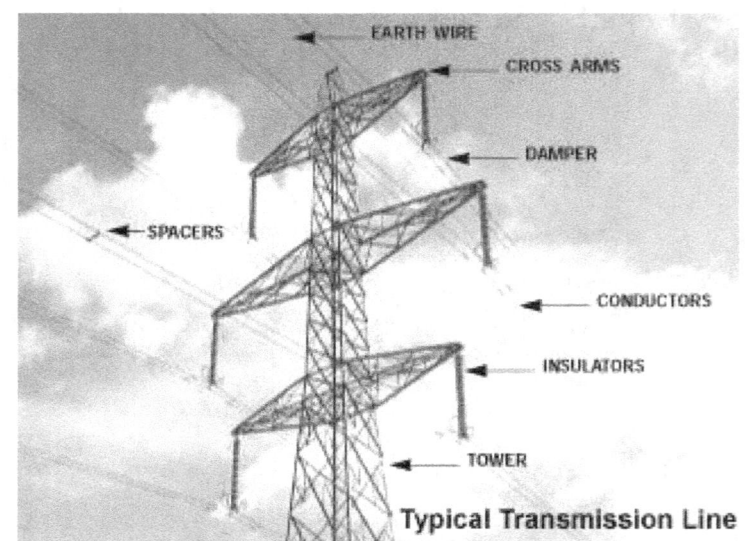

**Typical Transmission Line**

➢ Cable lines are designed to be placed underground or under water. The conductors are insulated from one another and surrounded by protective sheath. Cable lines are usually more expensive and harder to maintain. They also have capacitance problem – not suitable for long distance.

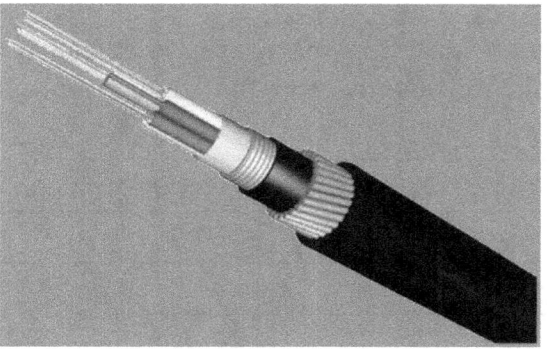

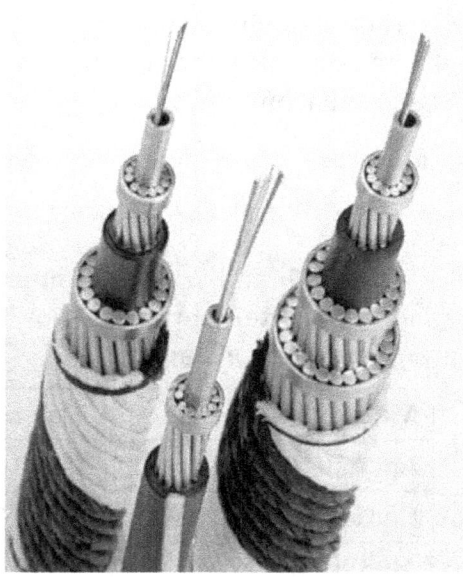

➢ Transmission lines are characterized by a series resistance, inductance and shunt capacitance per unit length. These values determine the power-carrying capacity of the transmission line and the voltage drop across it at full load.

# Resistance

The DC resistance of a conductor is given by

$$R_{DC} = \frac{\rho l}{A} \qquad (1)$$

Where *l* is the length of conductor; *A* – cross-sectional area, $\rho$ is the resistivity of the conductor. Therefore, the DC resistance per meter of the conductor is

$$r_{DC} = \frac{\rho}{A} \left[ \frac{\Omega}{m} \right] \qquad (2)$$

➢ The resistivity of a conductor is a fundamental property of the material that the conductor is made from. It varies with both type and temperature of the material. At the same temperature, the resistivity of aluminum is higher than the resistivity of copper.

➢ The resistivity increases linearly with temperature over normal range of temperatures. If the resistivity at one temperature is known, the resistivity at another temperature can be found from

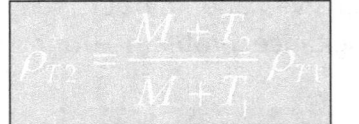

$$\rho_{T2} = \frac{M + T_2}{M + T_1} \rho_{T1}$$

(3)

➢ Where $T_1$ and $\rho_{T1}$ are temperature 1 in °C and the resistivity at that temperature, $T_2$ and $\rho_{T2}$ are temperature 2 in °C and the resistivity at that temperature, and $M$ is the temperature constant.

| Material | Resistivity at 20°C [Ω·m] | Temperature constant [°C] |
|---|---|---|
| Annealed copper | $1.72 \cdot 10^{-8}$ | 234.5 |
| Hard-drawn copper | $1.77 \cdot 10^{-8}$ | 241.5 |
| Aluminum | $2.83 \cdot 10^{-8}$ | 228.1 |
| Iron | $10.00 \cdot 10^{-8}$ | 180.0 |
| Silver | $1.59 \cdot 10^{-8}$ | 243.0 |

➢ We notice that silver and copper would be among the best conductors. However, aluminum, being much cheaper and lighter, is used to make most of the transmission line conductors. Conductors made out of aluminum should have bigger diameter than copper conductors to offset the higher resistivity of the material and, therefore, support the necessary currents.

➢ AC resistance of a conductor is always higher than its DC resistance due to the skin effect forcing more current flow near the outer surface of the conductor. The higher the frequency of current, the more noticeable skin effect would be.

➢ At frequencies of our interest (50-60 Hz), however, skin effect is not very strong.

➢ Wire manufacturers usually supply tables of resistance per unit length at common frequencies (50 and 60 Hz). Therefore, the resistance can be determined from such tables.

# TL Inductance and Inductive Reactance

LO4

# Inductance and Inductive Reactance

➤ The series inductance of a transmission line consists of two components: internal and external inductances, which are due the magnetic flux inside and outside the conductor respectively. The inductance of a transmission line is defined as the number of flux linkages [Wb-turns] produced per ampere of current flowing through the line:

$$L = \frac{\lambda}{I} \qquad (4)$$

## 1. Internal inductance:

> The total internal flux linkages per meter can be found via integration...

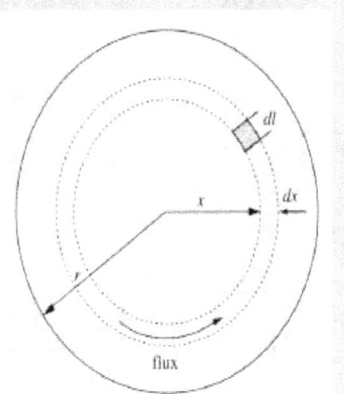

$$\lambda_{int} = \int d\lambda = \int_0^r \frac{\mu x^3 I}{2\pi r^4} dx = \frac{\mu I}{8\pi} \quad [Wb-turns/m] \tag{12}$$

> Therefore, the internal inductance per meter is

$$l_{int} = \frac{\lambda_{int}}{I} = \frac{\mu}{8\pi} \quad [H/m] \tag{13}$$

> If the relative permeability of the conductor is 1 (non-ferromagnetic materials, such as copper and aluminum), the inductance per meter reduces to

$$l_{int} = \frac{\mu_0}{8\pi} = \frac{4\pi \cdot 10^{-7}}{8\pi} = 0.5 \cdot 10^{-7} \quad [H/m] \tag{14}$$

## External inductance between two points outside of the line

> To find the inductance external to a conductor, we need to calculate the flux linkages of the conductor due only the portion of flux between two points $P_1$ and $P_2$ that lie at distances $D_1$ and $D_2$ from the center of the conductor.

The total external flux linkages per meter can be found via integration...

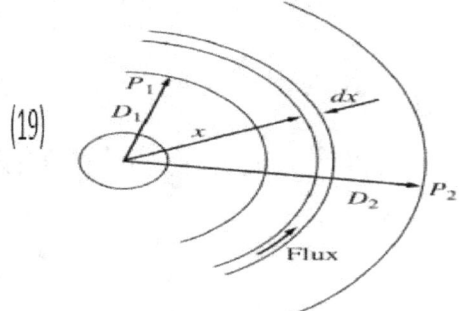

$$\lambda_{ext} = \int_{D_1}^{D_2} d\lambda = \int_{D_2}^{D_2} \frac{\mu I}{2\pi x} dx = \frac{\mu I}{2\pi} \ln \frac{D_2}{D_1} \quad [Wb-turns/m] \tag{19}$$

The external inductance per meter is

$$l_{ext} = \frac{\lambda_{ext}}{I} = \frac{\mu}{2\pi} \ln \frac{D_2}{D_1} \quad [H/m] \tag{20}$$

# Inductance of a single-phase two-wire transmission line

➤ We determine next the series inductance of a single-phase line consisting of two conductors of radii $r$ spaced by a distance $D$ and both carrying currents of magnitude $I$ flowing into the page in the left-hand conductor and out of the page in the right-hand conductor.

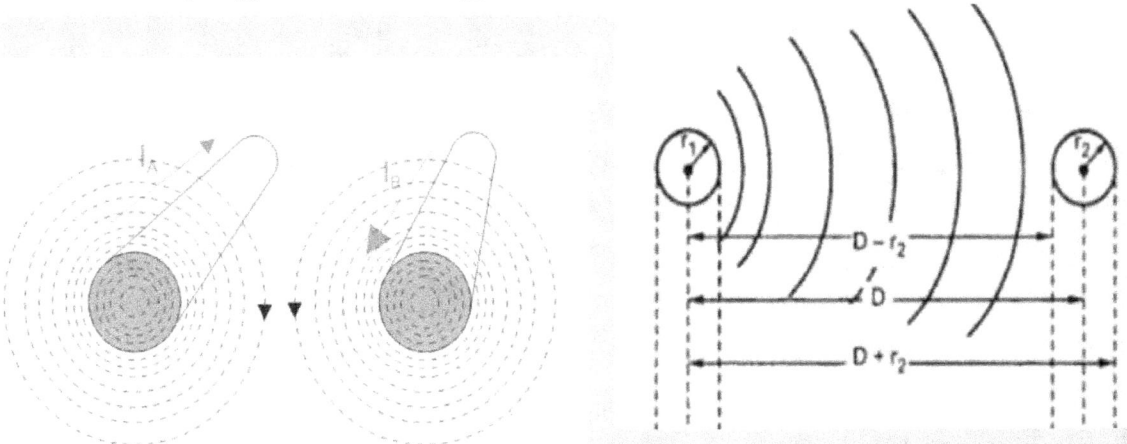

➤ The total inductance of a wire per unit length in this transmission line is a sum of the internal inductance and the external inductance between the conductor surface ($r$) and the separation distance ($D$):

$$l = l_{int} + l_{ext} = \frac{\mu}{2\pi}\left(\frac{1}{4} + \ln\frac{D}{r}\right) \quad [H/m] \tag{22}$$

➤ By symmetry, the total inductance of the other wire is the same, therefore, the total inductance of a two-wire transmission line is

$$l = \frac{\mu}{\pi}\left(\frac{1}{4} + \ln\frac{D}{r}\right) \quad [H/m] \tag{23}$$

➤ Where $r$ is the radius of each conductor and $D$ is the distance between conductors.

# Inductance of a transmission line

➤ Equations similar to (23) can be derived for three-phase lines and for lines with more phases... In most of the practical situations, the inductance of the transmission line can be found from tables supplied by line developers.

Analysis of (23) shows that:

1. The greater the spacing between the phases of a transmission line, the greater the inductance of the line. Since the phases of a high-voltage overhead transmission line must be spaced further apart to ensure proper insulation, a high-voltage line will have a higher inductance than a low-voltage line. Since the spacing between lines in buried cables is very small, series inductance of cables is much smaller than the inductance of overhead lines.

2. The greater the radius of the conductors in a transmission line, the lower the inductance of the line. In practical transmission lines, instead of using heavy and inflexible conductors of large radii, two and more conductors are bundled together to approximate a large diameter conductor. The more conductors included in the bundle, the better the approximation becomes. Bundles are often used in the high-voltage transmission lines.

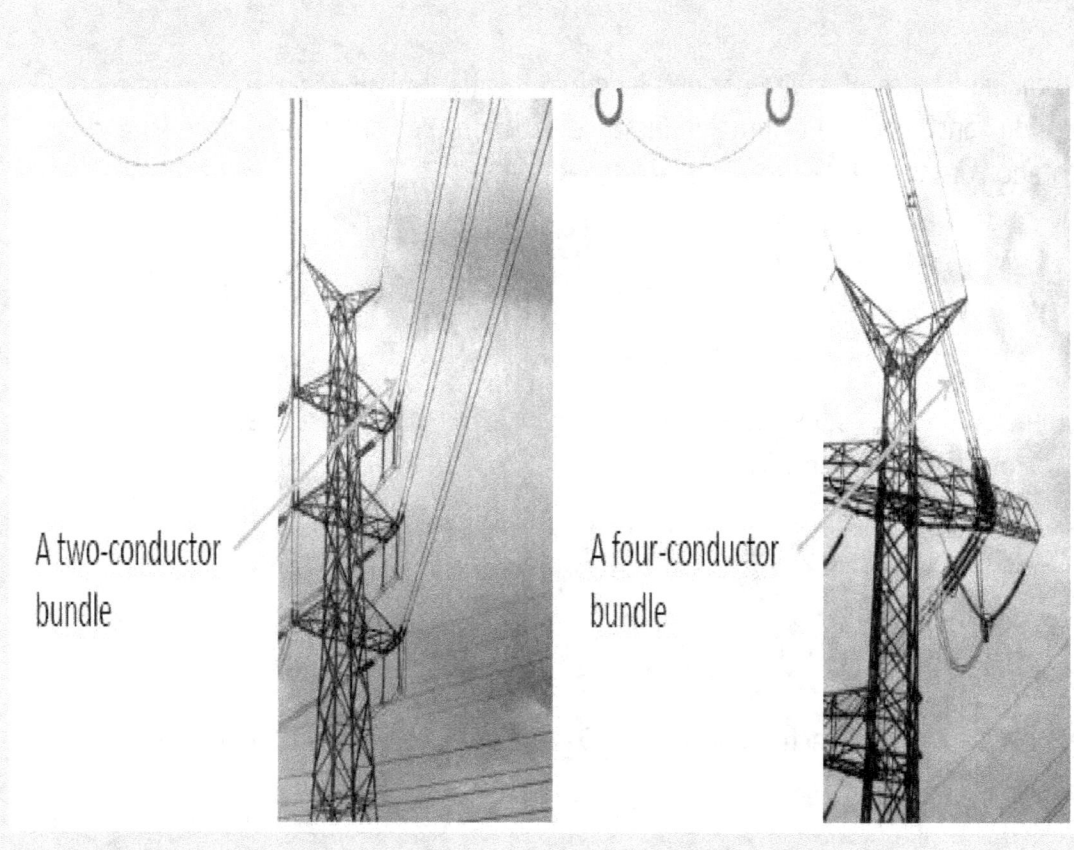

A two-conductor bundle

A four-conductor bundle

# Inductive reactance of a line

➤ The series inductive reactance of a transmission line depends on both the inductance of the line and the frequency of the power system. Denoting the inductance per unit length as $l$, the inductive reactance per unit length will be

$$x_I = j\omega l = j2\pi f l \qquad (24)$$

➤ where $f$ is the power system frequency. Therefore, the total series inductive reactance of a transmission line can be found as

$$X_I = x_I d \qquad (25)$$

➤ where $d$ is the length of the line.

# Inductance Simplification

Inductance expression can be simplified using two exponential identities:

$$\ln(ab) = \ln a + \ln b \qquad \ln \frac{a}{b} = \ln a - \ln b \qquad a = \ln(e^a)$$

$$l = \frac{\mu}{\pi}\left(\frac{1}{4} + \ln \frac{D}{r}\right) = \frac{\mu}{\pi}\left(\ln e^{\frac{1}{4}} + \ln \frac{D}{r}\right) = \frac{\mu}{\pi}\left(\ln \frac{D}{r e^{\frac{-1}{4}}}\right)$$

$$\mu = \mu_r \mu_0 \qquad \therefore \mu_o = 4\pi \times 10^{-7}$$

➤ For air the $\mu_r=1$

$$l = \frac{\mu}{\pi}\left(\ln \frac{D}{re^{\frac{-1}{4}}}\right) = 4 \times 10^{-7} \ln \frac{D}{re^{\frac{-1}{4}}}$$

$$\because r' = re^{\frac{-1}{4}}$$

$$\therefore l = 4 \times 10^{-7} \ln \frac{D}{r'} \quad [H/m]$$

➤ The value of the inductance is called _inductance per loop meter_ or _per loop mile_ to distinguish.

## Many-Conductor Case

Now assume we now have _k_ conductors, each with current $i_k$, arranged in some specified geometry. We'd like to find flux linkages of each conductor.

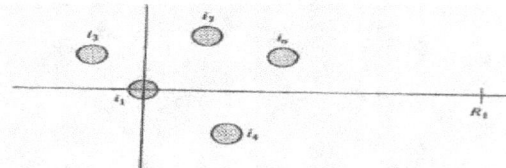

Each conductor's flux linkage, $\lambda_k$, depends upon its own current and the current in all the other conductors.

$$\lambda_1 = \frac{\mu_0}{2\pi}\left[i_1 \ln \frac{1}{r_1'} + i_2 \ln \frac{1}{d_{12}} + \cdots + i_n \ln \frac{1}{d_{1n}}\right],$$

$$= L_{11}i_1 + L_{12}i_2 \cdots + L_{1n}i_n$$

# Line Inductance Example

Calculate the reactance for a balanced 3$\phi$, 60Hz transmission line with a conductor geometry of an equilateral triangle with $D$ = 5m, $r$ = 1.24cm (Rook conductor) and a length of 5 miles.

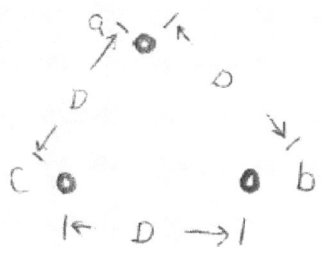

Since system is assumed balanced

$$i_a = -i_b - i_c$$

$$\lambda_a = \frac{\mu_0}{2\pi}\left[i_a \ln(\frac{1}{r'}) + i_b \ln(\frac{1}{D}) + i_c \ln(\frac{1}{D})\right]$$

Substituting $i_a = -i_b - i_c$, obtain:

$$\lambda_a = \frac{\mu_0}{2\pi}\left[i_a \ln\left(\frac{1}{r'}\right) - i_a \ln\left(\frac{1}{D}\right)\right]$$

$$= \frac{\mu_0}{2\pi} i_a \ln\left(\frac{D}{r'}\right).$$

$$L_a = \frac{\mu_0}{2\pi}\ln\left(\frac{D}{r'}\right) = \frac{4\pi \times 10^{-7}}{2\pi}\ln\left(\frac{5}{9.67 \times 10^{-3}}\right)$$

$$= 1.25 \times 10^{-6} \text{ H/m}.$$

Again note logarithm of ratio of distance between phases to the size of the conductor.

$$L_a = 1.25 \times 10^{-6} \text{ H/m}$$

Converting to reactance

$$X_a = 2\pi \times 60 \times 1.25 \times 10^{-6}$$

$$= 4.71 \times 10^{-4} \ \Omega/\text{m}$$

$$= 0.768 \ \Omega/\text{mile}$$

$$X_{\text{Total for 5 mile line}} = 3.79 \ \Omega$$

(this is the total per phase)

The reason we did NOT have mutual inductance was because of the symmetric conductor spacing

# Bundled Conductor Flux Linkages

➤For the line shown on the left, define $d_{ij}$ as the distance between conductors $i$ and $j$.

➤ We can then determine $l_k$ for conductor $k$.

➤Assuming ¼ of the phase current flows in each of the four conductors in a given phase bundle, then for conductor 1:

$$\lambda_1 = \frac{\mu_0}{2\pi} \left[ \begin{array}{l} \dfrac{i_a}{4}\left( \ln\dfrac{1}{r'} + \ln\dfrac{1}{d_{12}} + \ln\dfrac{1}{d_{13}} + \ln\dfrac{1}{d_{14}} \right) + \\[2ex] \dfrac{i_b}{4}\left( \ln\dfrac{1}{d_{15}} + \ln\dfrac{1}{d_{16}} + \ln\dfrac{1}{d_{17}} + \ln\dfrac{1}{d_{18}} \right) + \\[2ex] \dfrac{i_c}{4}\left( \ln\dfrac{1}{d_{19}} + \ln\dfrac{1}{d_{1,10}} + \ln\dfrac{1}{d_{1,11}} + \ln\dfrac{1}{d_{1,12}} \right) \end{array} \right]$$

If $D_{ab} = D_{ac} = D_{bc} = D$ and $i_a = -i_b - i_c$
Then

$$\lambda_1 = \frac{\mu_0}{2\pi}\left[i_a \ln\left(\frac{1}{R_b}\right) - i_a \ln\left(\frac{1}{D}\right)\right]$$

$$= \frac{\mu_0}{2\pi} I_a \ln\left(\frac{D}{R_b}\right) = \frac{\mu_0}{2\pi} 4I_1 \ln\left(\frac{D}{R_b}\right)$$

$$L_1 = \frac{\mu_0}{2\pi} \times 4 \times \ln\left(\frac{D}{R_b}\right), \text{ which is the}$$

self-inductance of wire 1.

$R_b$ = geometric mean radius (GMR) of bundle

$$= (r'd_{12}d_{13}d_{14})^{1/4} \text{ for our example}$$

$$= (r'd_{12}\ldots d_{1b})^{1/b} \text{ in general}$$

But remember each bundle has $b$ conductors in parallel (4 in this example).

So, there are four inductances in parallel:

$$L_a = L_1/b = \frac{\mu_0}{2\pi} \ln\left(\frac{D}{R_b}\right).$$

Again note that inductance depends on the logarithm of the ratio of distance between phases to the size of bundle of conductors.

Inductance decreases with decreasing distance between phases and increasing bundle size.

# Bundle Inductance Example

Consider the previous example of the three phases symmetrically spaced 5 meters apart using wire with a radius of $r$ = 1.24 cm. Except now assume each phase has 4 conductors in a square bundle, spaced 0.25 meters apart. What is the new inductance per meter?

$$r = 1.24 \times 10^{-2} \text{ m} \quad r' = 9.67 \times 10^{-3} \text{ m}$$

0.25 M

0.25 M          0.25 M

$$R_b = \left(9.67 \times 10^{-3} \times 0.25 \times 0.25 \times (\sqrt{2} \times 0.25)\right)^{1/4}$$

$$= 0.12 \text{ m} \quad (\text{ten times bigger than } r!)$$

$$L_a = \frac{\mu_0}{2\pi} \ln \frac{5}{0.12} = 7.46 \times 10^{-7} \text{ H/m}$$

Bundling reduces inductance.

# Transmission Tower Configurations

➤ The problem with the line analysis we've done so far is we have assumed a symmetrical tower configuration.

➤ Such a tower configuration is seldom practical.

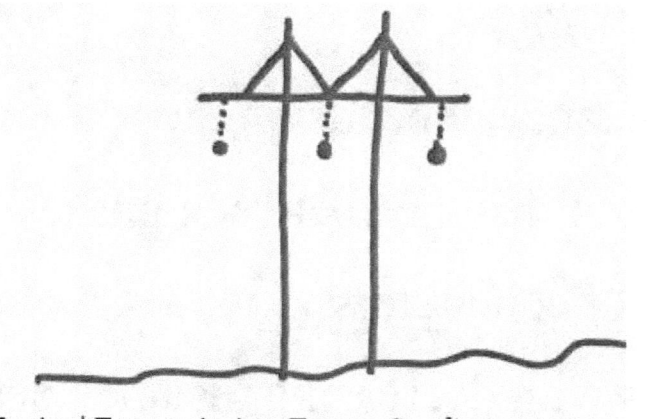

Typical Transmission Tower Configuration

➤ Therefore in general $D_{ab} \neq D_{ac} \neq D_{bc}$

➤ Unless something was done this would result in unbalanced Phases.

# Transposition

- To keep system balanced, over the length of a transmission line the conductors are "rotated" so each phase occupies each position on tower for an equal distance.
- This is known as transposition.

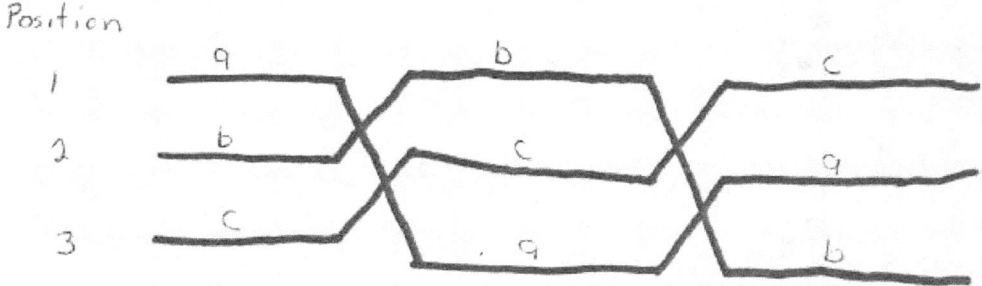

Aerial or side view of conductor positions over the length
of the transmission line.

# Line Transposition Example

For a uniformly transposed line we can calculate the flux linkage for phase "a"

$$\lambda_a = \frac{1}{3}\frac{\mu_0}{2\pi}\left[I_a \ln\frac{1}{r'} + I_b \ln\frac{1}{d_{12}} + I_c \ln\frac{1}{d_{13}}\right] +$$ "a" phase in position "1"

$$\frac{1}{3}\frac{\mu_0}{2\pi}\left[I_a \ln\frac{1}{r'} + I_b \ln\frac{1}{d_{13}} + I_c \ln\frac{1}{d_{23}}\right] +$$ "a" phase in position "3"

$$\frac{1}{3}\frac{\mu_0}{2\pi}\left[I_a \ln\frac{1}{r'} + I_b \ln\frac{1}{d_{23}} + I_c \ln\frac{1}{d_{12}}\right]$$ "a" phase in position "2"

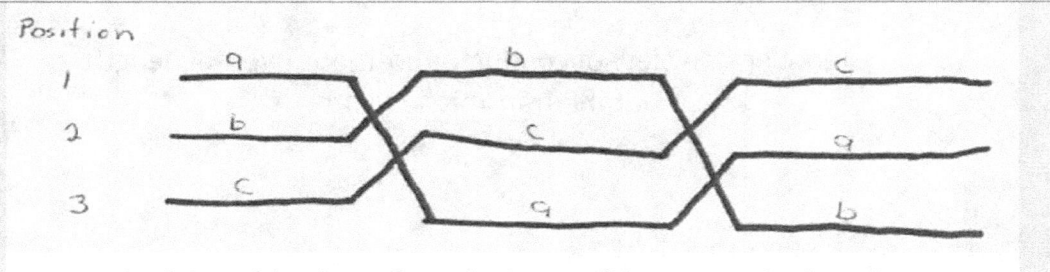

Recognizing that

$$\frac{1}{3}(\ln a + \ln b + \ln c) = \ln(abc)^{1/3}$$

We can simplify so

$$\lambda_a = \frac{\mu_0}{2\pi}\left[\begin{array}{l} I_a \ln\frac{1}{r'} + I_b \ln\frac{1}{(d_{12}d_{13}d_{23})^{1/3}} + \\ I_c \ln\frac{1}{(d_{12}d_{13}d_{23})^{1/3}} \end{array}\right]$$

## *Inductance of Transposed Line*

Define the geometric mean distance (GMD)

$$D_m = (d_{12}d_{13}d_{23})^{1/3}$$

Then for a balanced $3\phi$ system $(I_a = -I_b - I_c)$

$$\lambda_a = \frac{\mu_0}{2\pi}\left[I_a \ln\frac{1}{r'} - I_a \ln\frac{1}{D_m}\right] = \frac{\mu_0}{2\pi}I_a \ln\frac{D_m}{r'}$$

Hence

$$L_a = \frac{\mu_0}{2\pi}\ln\frac{D_m}{r'} = 2\times10^{-7}\ln\frac{D_m}{r'} \text{ H/m}$$

# Inductance with Bundling

If the line is bundled with a geometric mean radius, $R_b$, then

$$\lambda_a = \frac{\mu_0}{2\pi} I_a \ln \frac{D_m}{R_b}$$

$$L_a = \frac{\mu_0}{2\pi} \ln \frac{D_m}{R_b} = 2 \times 10^{-7} \ln \frac{D_m}{R_b} \quad \text{H/m}$$

# Inductance with Bundling

If the line is bundled with a geometric mean radius, $R_b$, then

$$\lambda_a = \frac{\mu_0}{2\pi} I_a \ln \frac{D_m}{R_b}$$

$$L_a = \frac{\mu_0}{2\pi} \ln \frac{D_m}{R_b} = 2 \times 10^{-7} \ln \frac{D_m}{R_b} \quad \text{H/m}$$

# Inductance Example

> Calculate the per phase inductance and reactance of a balanced 3φ, 60 Hz, line with:

 - horizontal phase spacing of 10m
 - using three conductor bundling with a spacing between conductors in the bundle of 0.3m.

> Assume the line is uniformly transposed and the conductors have a 1cm radius.

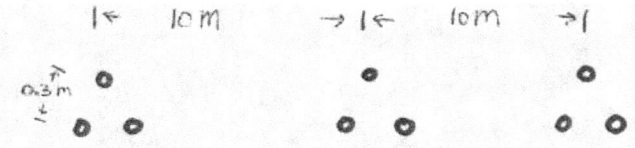

$$D_m = (d_{12}d_{13}d_{23})^{\frac{1}{3}},$$

$$= (10 \times (2 \times 10) \times 10)^{1/3} = 12.6\text{m},$$

$$r' = re^{-\mu_r/4} = 0.0078\text{m},$$

$$R_b = (r'd_{12}d_{13})^{\frac{1}{3}},$$

$$= (r' \times 0.3 \times 0.3)^{1/3} = 0.0888\text{m},$$

$$L_a = \frac{\mu_0}{2\pi}\ln\frac{D_m}{R_b}$$

$$= 9.9 \times 10^{-7}\,\text{H/m},$$

$$X_a = 2\pi f L_a (1600\text{m/mile}) = 0.6\,\Omega/\text{mile}.$$

**Example:** Find GMD, GMR for each circuit, inductance for each circuit, and total inductance per meter for two circuits that run parallel to each other. One circuit consists of three 0.25 cm radius conductors. The second circuit consists of two 0.5 cm radius conductor

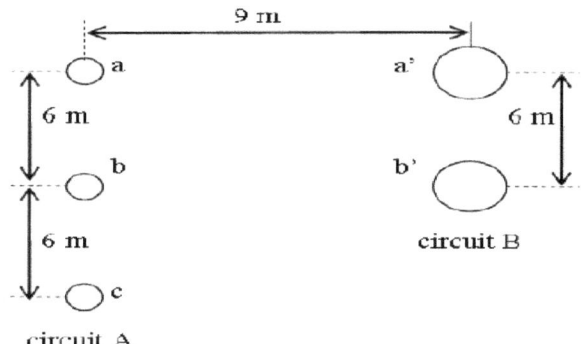

**Solution:**

$m = 3$, $n' = 2$, $\therefore m \cdot n' = 6$

$$GMD = \sqrt[6]{D_{aa'} \cdot D_{ab'} \cdot D_{ba'} \cdot B_{bb'} \cdot D_{ca'} \cdot D_{cb'}}$$

where

$D_{aa'} = D_{bb'} = 9\ m$

$D_{ab'} = D_{ba'} = D_{cb'} = \sqrt{6^2 + 9^2} = \sqrt{117}\ m$

$D_{ca'} = \sqrt{12^2 + 9^2} = 15\ m$

$\therefore GMD = 10.743\ m$

Geometric Mean Radius for Circuit A:

$$GMR_A = \sqrt[3^2]{D_{aa}D_{ab}D_{ac}D_{ba}D_{bb}D_{bc}D_{ca}D_{cb}D_{cc}} = \sqrt[9]{\left(0.25 \times 10^{-2} \times e^{-\frac{1}{4}}\right)^3 \times 6^4 \times 12^2} = 0.481 m$$

Geometric Mean Radius for Circuit B:

$$GMR_B = \sqrt[2^2]{D_{a'a'}D_{a'b'}D_{b'b'}D_{b'a'}} = \sqrt[4]{\left(0.5 \times 10^{-2} \times e^{-\frac{1}{4}}\right)^2 \times 6^2} = 0.153 m$$

Inductance of circuit A

$$L_A = 2 \times 10^{-7} \ln \frac{GMD}{GMR_A} = 2 \times 10^{-7} \ln \frac{10.743}{0.481} = 6.212 \times 10^{-7} \qquad H/m$$

Inductance of circuit B

$$L_B = 2 \times 10^{-7} \ln \frac{GMD}{GMR_B} = 2 \times 10^{-7} \ln \frac{10.743}{0.153} = 8.503 \times 10^{-7} \qquad H/m$$

The total inductance is then

$$L_T = L_A + L_B = 14.715 \times 10^{-7} \qquad H/m$$

# Example

An 8000 V, 60 Hz, single-phase, transmission line consists of two hard-drawn aluminum conductors with a radius of 2 cm spaced 1.2 m apart. If the transmission line is 30 km long and the temperature of the conductors is 20°C,

a.  What is the series resistance per kilometer of this line?
b.  What is the series inductance per kilometer of this line?
c.  What is the total series impedance of this line?

a. The series resistance of the transmission line is

$$R = \frac{\rho l}{A}$$

Ignoring the skin effect, the resistivity of the line at $20^0$ will be $2.83 \cdot 10^{-8}$ Ω-m and the resistance per kilometer of the line is

$$r = \frac{\rho l}{A} = \frac{2.83 \cdot 10^{-8} \cdot 1000}{\pi \cdot 0.02^2} = 0.0225 \ \Omega/km$$

b. The series inductance per kilometer of the transmission line is

$$l = \frac{\mu}{\pi}\left(\frac{1}{4} + \ln\frac{D}{r}\right) \cdot 1000 = \frac{\mu}{\pi}\left(\frac{1}{4} + \ln\frac{1.2}{0.02}\right) \cdot 1000 = 1.738 \cdot 10^{-3} \ H/km$$

d. The series impedance per kilometer of the transmission line is

$$z_{se} = r + jx = r + j2\pi fl = 0.0225 + j2\pi \cdot 60 \cdot 1.738 \cdot 10^{-3} = 0.0225 + j0.655 \ \Omega/km$$

Then the total series impedance of the line is

$$Z_{se} = (0.0225 + j0.655) \cdot 30 = 0.675 + j19.7 \ \Omega$$

HIGHER COLLEGES OF TECHNOLOGY

# TL Capacitance and Capacitive Reactance

# Capacitance and Capacitive Reactance

➢ Since a voltage $V$ is applied to a pair of conductors separated by a dielectric (air), charges of equal magnitude but opposite sign will accumulate on the conductors:

$$q = CV \qquad (1)$$

➢ Where $C$ is the capacitance between the pair of conductors.

➢ In AC power systems, a transmission line carries a time-varying voltage different in each phase. This time-varying voltage causes the changes in charges stored on conductors. Changing charges produce a changing current, which will increase the current through the transmission line and affect the power factor and voltage drop of the line. This changing current will flow in a transmission line even if it is open circuited.

➢ The capacitance of the transmission line can be found using the Gauss's law:

$$\oiint_A D \cdot dA = q \qquad (2)$$

➢ where $A$ specifies a closed surface; $dA$ is the unit vector normal to the surface; $q$ is the charge inside the surface; $D$ is the electric flux density at the surface:

$$D = \varepsilon E \qquad (3)$$

➢ where $E$ is the electric field intensity at that point; $\varepsilon$ is the permittivity of the material:

$$\varepsilon = \varepsilon_r \varepsilon_0 \qquad (4)$$

Relative permittivity of the material

➢ The permittivity of free space $\varepsilon_0 = 8.85 \cdot 10^{-12}$ F/m

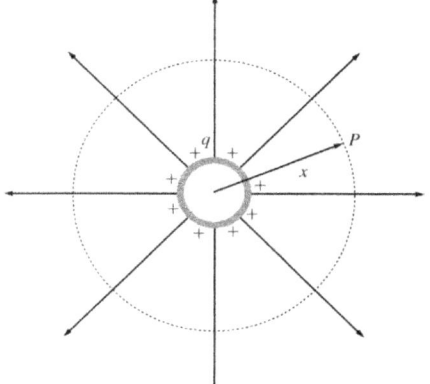

➤ Electric flux lines radiate uniformly outwards from the surface of the conductor with a positive charge on its surface. In this case, the flux density vector $D$ is always parallel to the normal vector $dA$ and is constant at all points around a path of constant radius $r$. Therefore:

$$DA = Q \implies D(2\pi x l) = ql \qquad (5)$$

➤ where $l$ is the length of conductor; $q$ is the charge density; $Q$ is the total charge on the conductor.

Then the flux density is
$$D = \frac{q}{2\pi x} \qquad (6)$$

The electric field intensity is
$$E = \frac{q}{2\pi \varepsilon x} \qquad (7)$$

➤ The potential difference between two points $P_1$ and $P_2$ can be found as

$$V_{12} = \int_{P_1}^{P_2} E \cdot dl \qquad (8)$$

➤ where $dl$ is a differential element tangential to the integration path between $P_1$ and $P_2$. The path is irrelevant.

➤ Selection of path can simplify calculations.
For $P_1$ - $P_{int}$, vectors $E$ and $dl$ are parallel; therefore, $E \cdot dl = Edx$. For $P_{int} - P_2$ vectors are orthogonal, therefore $E \cdot dl = 0$.

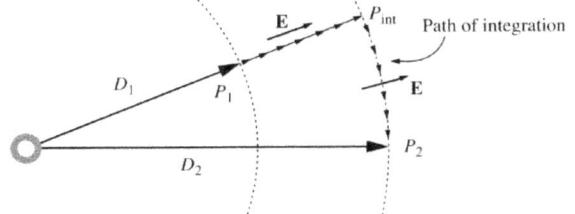

$$V_{12} = \int_{D_1}^{D_2} Edx = \int_{D_1}^{D_2} \frac{q}{2\pi \varepsilon x} dx = \frac{q}{2\pi \varepsilon} \ln \frac{D_2}{D_1} \qquad (9)$$

  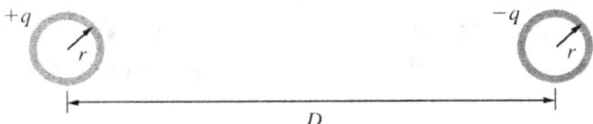
# Capacitance of a Single Phase Two-wire Transmission Line

➤ The potential difference due to the charge on conductor *a* can be found as

$$V_{ab,a} = \frac{q_a}{2\pi\varepsilon} \ln\frac{D}{r} \tag{10}$$

Similarly, the potential difference due to the charge on conductor *b* is

$$V_{ba,b} = \frac{q_b}{2\pi\varepsilon} \ln\frac{D}{r} \tag{11}$$

or

$$V_{ab,b} = -\frac{q_b}{2\pi\varepsilon} \ln\frac{D}{r} \tag{12}$$

➤ The total voltage between the lines is

$$V_{ab} = V_{ab,a} - V_{ab,b} = \frac{q_a}{2\pi\varepsilon} \ln\frac{D}{r} - \frac{-q_b}{2\pi\varepsilon} \ln\frac{D}{r} \tag{13}$$

➤ Since $q_1 = q_2 = q$, the equation reduces to

$$V_{ab} = \frac{q}{\pi\varepsilon} \ln\frac{D}{r} \tag{14}$$

➤ The capacitance per unit length between the two conductors of the line is

$$C_{ab} = \frac{q}{V} = \frac{q}{\dfrac{q}{\pi\varepsilon} \ln\dfrac{D}{r}} \tag{15}$$

Thus:

$$c_{ab} = \frac{\pi \varepsilon}{\ln \dfrac{D}{r}}$$

(16)

Which is the capacitance per unit length of a single-phase two-wire transmission line.

The potential difference between each conductor and the ground (or neutral) is one half of the potential difference between the two conductors. Therefore, the capacitance to ground of this single-phase transmission line will be

$$c_n = c_{an} = c_{bn} = \frac{2\pi \varepsilon}{\ln \dfrac{D}{r}}$$

(17)

➤ Similarly, the expressions for capacitance of three-phase lines (and for lines with more than 3 phases) can be derived. Similarly to the inductance, the capacitance of the transmission line can be found from tables supplied by line developers.

1. The greater the spacing between the phases of a transmission line, the lower the capacitance of the line. Since the phases of a high-voltage overhead transmission line must be spaced further apart to ensure proper insulation, a high-voltage line will have a lower capacitance than a low-voltage line. Since the spacing between lines in buried cables is very small, shunt capacitance of cables is much larger than the capacitance of overhead lines. Cable lines are normally used for short transmission lines (to min capacitance) in urban areas.
2. The greater the radius of the conductors in a transmission line, the higher the capacitance of the line. Therefore, bundling increases the capacitance. Good transmission line is a compromise among the requirements for low series inductance, low shunt capacitance, and a large enough separation to provide insulation between the phases.

# Shunt capacitive admittance

The shunt capacitive admittance of a transmission line depends on both the capacitance of the line and the frequency of the power system. Denoting the capacitance per unit length as *c*, the shunt admittance per unit length will be

$$y_c = j\omega c = j2\pi fc \qquad (18)$$

The total shunt capacitive admittance therefore is

$$Y_c = y_c d = j2\pi fcd \qquad (20)$$

where *d* is the length of the line. The corresponding capacitive reactance is the reciprocal to the admittance:

$$Z_c = \frac{1}{Y_c} = -j\frac{1}{2\pi fcd} \qquad (21)$$

## Example

An 8000 V, 60 Hz, single-phase, transmission line consists of two hard-drawn aluminum conductors with a radius of 2 cm spaced 1.2 m apart. If the transmission line is 30 km long and the temperature of the conductors is 20°C,
a.   What is the series resistance per kilometer of this line?
b.   What is the series inductance per kilometer of this line?
c.   What is the shunt capacitance per kilometer of this line?
d.   What is the total series impedance of this line?
e.   What is the total shunt admittance of this line?

a. The series resistance of the transmission line is $\quad R = \dfrac{\rho l}{A}$

Ignoring the skin effect, the resistivity of the line at 20° will be 2.83·10⁻⁸ Ω-m and the resistance per kilometer of the line is

$$r = \frac{\rho l}{A} = \frac{2.83 \cdot 10^{-8} \cdot 1000}{\pi \cdot 0.02^2} = 0.0225 \ \Omega/km$$

b. The series inductance per kilometer of the transmission line is

$$l = \frac{\mu}{\pi}\left(\frac{1}{4} + \ln\frac{D}{r}\right)\cdot 1000 = \frac{\mu}{\pi}\left(\frac{1}{4} + \ln\frac{1.2}{0.02}\right)\cdot 1000 = 1.738\cdot 10^{-3}\ \ H/km$$

c. The shunt capacitance per kilometer of the transmission line is

$$c_{ab} = \frac{\pi\varepsilon}{\ln\dfrac{D}{r}}\cdot 1000 = \frac{\pi\cdot 8.854\cdot 10^{-12}}{\ln\dfrac{1.2}{0.02}}\cdot 1000 = 6.794\cdot 10^{-9}\ \ F/km$$

d. The series impedance per kilometer of the transmission line is

$$z_{se} = r + jx = r + j2\pi fl = 0.0225 + j2\pi\cdot 60\cdot 1.738\cdot 10^{-3} = 0.0225 + j0.655\ \ \Omega/km$$

Then the total series impedance of the line is

$$Z_{se} = (0.0225 + j0.655)\cdot 30 = 0.675 + j19.7\ \ \Omega$$

e. The shunt admittance per kilometer of the transmission line is

$$y_C = j2\pi fc = j2\pi\cdot 60\cdot 6.794\cdot 10^{-9} = j2.561\cdot 10^{-6}\ \ S/m$$

The total shunt admittance will be

$$Y_{se} = (j2.561\cdot 10^{-6})\cdot 30 = j7.684\cdot 10^{-5}\ \ S$$

The corresponding shunt capacitive reactance is

$$Z_{sh} = \frac{1}{Y_{sh}} = \frac{1}{j7.684\cdot 10^{-5}} = -j13.0\ \ k\Omega$$

# Three Conductor Case

A

C                    B

Assume we have three infinitely long conductors, A, B, & C, each with radius r and distance D from each other.

Assume charge densities such that
$$q_a + q_b + q_c = 0$$

$$V_a = \frac{1}{2\pi\varepsilon}\left[ q_a \ln\frac{1}{r} + q_b \ln\frac{1}{D} + q_c \ln\frac{1}{D} \right]$$

$$V_a = \frac{q_a}{2\pi\varepsilon} \ln\frac{D}{r}$$

# Line Capacitance

For a single capacitor, capacitance is defined as

$$q_i = C_i V_i$$

But for a multiple conductor case we need to use matrix relationships since the charge on conductor $i$ may be a function of $V_j$

$$\begin{bmatrix} q_1 \\ \vdots \\ q_n \end{bmatrix} = \begin{bmatrix} C_{11} & \cdots & C_{1n} \\ \vdots & \cdots & \vdots \\ C_{n1} & \cdots & C_{nn} \end{bmatrix} \begin{bmatrix} V_1 \\ \vdots \\ V_n \end{bmatrix}$$

$$\mathbf{q} = \mathbf{C}\,\mathbf{V}$$

We will not be considering the cases with mutual capacitance. To eliminate mutual capacitance we'll again assume we have a uniformly transposed line, using similar arguments to the case of inductance. For the previous three conductor example:

Since $q_a = C V_a \implies C = \dfrac{q_a}{V_a} = \dfrac{2\pi\varepsilon}{\ln D/r}$

## *Bundled Conductor Capacitance*

Similar to the case for determining line inductance when there are $n$ bundled conductors, we use the original capacitance equation just substituting an equivalent radius

$$R_b^c = (r d_{12} \cdots d_{1n})^{1/n}$$

Note for the capacitance equation we use $r$ rather than $r'$ which was used for $R_b$ in the inductance equation

For the case of uniformly transposed lines we use the same GMR, $D_m$, as before.

$$C = \frac{2\pi\varepsilon}{\ln \dfrac{D_m}{R_b^c}}$$

where

$$D_m = \left[d_{ab}d_{ac}d_{bc}\right]^{1/3}$$

$$R_b^c = (rd_{12}\cdots d_{1n})^{1/n} \quad (\text{note } r \text{ NOT } r')$$

$$\varepsilon \text{ in air } = \varepsilon_o = 8.854\times10^{-12} \text{ F/m}$$

## Example

*Calculate the per phase capacitance and susceptance of a balanced 3$\phi$, 60 Hz, transmission line with horizontal phase spacing of 10m using three conductor bundling with a spacing between conductors in the bundle of 0.3m. Assume the line is uniformly transposed and the conductors have a 1cm radius.*

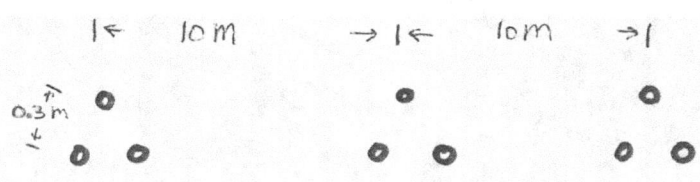

$$R_b^c = (0.01 \times 0.3 \times 0.3)^{1/3} = 0.0963 \text{ m}$$

$$D_m = (10 \times 10 \times 20)^{1/3} = 12.6 \text{ m}$$

$$C = \frac{2\pi \times 8.854 \times 10^{-12}}{\ln \dfrac{12.6}{0.0963}} = 1.141 \times 10^{-11} \text{ F/m}$$

$$X_c = \frac{1}{\omega C} = \frac{1}{2\pi 60 \times 1.141 \times 10^{-11} \text{ F/m}}$$

$$= 2.33 \times 10^8 \ \Omega\text{-m (not } \Omega/\text{m)}$$

# *Line Conductors*

➢ Typical transmission lines use multi-strand conductors

➢ ACSR (aluminum conductor steel reinforced) conductors are most common. A typical Al. to St. ratio is about 4 to 1.

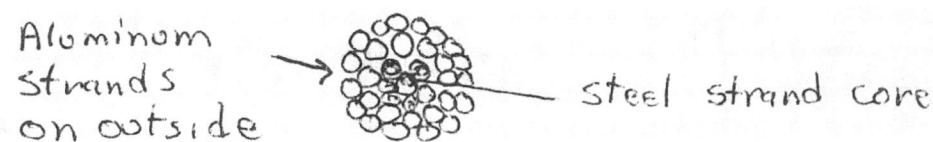

Aluminum strands on outside → steel strand core

> Total conductor area is given in circular mils. One circular mil is the area of a circle with a diameter of 0.001 inch, and so has an area of $\pi \times 0.0005^2$ square inches.

> **Example:** what is the area of a solid, 1" diameter circular wire?

**Answer:** 1000 kcmil (kilo circular mils)

> Because conductors are stranded, the equivalent radius must be provided by the manufacturer in tables. This value is known as the GMR and is usually expressed in feet.

# ACSR Table Data

**TABLE A8.1.** BARE ALUMINUM CONDUCTORS, STEEL REINFORCED (ACSR)
ELECTRICAL PROPERTIES OF MULTILAYER SIZES  *(Cont'd)*

| Code Word | Size (kcmil) | Stranding Al./St. | Number of Aluminum Layers | dc 20°C (Ohms/ Mile) | ac-60 Hz 25°C (Ohms/ Mile) | ac-60 Hz 50°C (Ohms/ Mile) | ac-60 Hz 75°C (Ohms/ Mile) | GMR (ft) | Inductive Ohms/ Mile $X_a$ | Capacitive Megohm-Miles $X'_a$ |
|---|---|---|---|---|---|---|---|---|---|---|
| Flicker | 477 | 24/7 | 2 | 0.1889 | 0.194 | 0.213 | 0.232 | 0.0283 | 0.432 | 0.0992 |
| Hawk | 477 | 26/7 | 2 | 0.1883 | 0.193 | 0.212 | 0.231 | 0.0290 | 0.430 | 0.0988 |
| Hen | 477 | 30/7 | 2 | 0.1869 | 0.191 | 0.210 | 0.229 | 0.0304 | 0.424 | 0.0980 |
| Osprey | 556.5 | 18/1 | 2 | 0.1629 | 0.168 | 0.184 | 0.200 | 0.0284 | 0.432 | 0.0981 |
| Parakeet | 556.5 | 24/7 | 2 | 0.1620 | 0.166 | 0.183 | 0.199 | 0.0306 | 0.423 | 0.0969 |
| Dove | 556.5 | 26/7 | 2 | 0.1613 | 0.166 | 0.182 | 0.198 | 0.0313 | 0.420 | 0.0965 |
| Eagle | 556.5 | 30/7 | 2 | 0.1602 | 0.164 | 0.180 | 0.196 | 0.0328 | 0.415 | 0.0957 |
| Peacock | 605 | 24/7 | 2 | 0.1490 | 0.153 | 0.168 | 0.183 | 0.0319 | 0.418 | 0.0957 |
| Squab | 605 | 26/7 | 2 | 0.1485 | 0.153 | 0.167 | 0.182 | 0.0327 | 0.415 | 0.0953 |

GMR is equivalent to $r'$     Inductance and Capacitance assuming a $D_m$ of 1 ft.

# ACSR Table Data

| Code word | Size, Mcmil | Stranding aluminum /steel | Outside diameter, in | Resistance | | GMR, ft | Phase-to-neutral, 60 Hz, reactance at 1-ft spacing | |
| --- | --- | --- | --- | --- | --- | --- | --- | --- |
| | | | | DC, Ω/ 1000 ft at 20°C | AC, 60 Hz, Ω/mi at 25°C | | Inductive Ω/mi, $X_a$ | Capacitive Ω/mi, $X_a$ |
| Waxwing | 266.8 | 18/1 | 0.609 | 0.0646 | 0.3448 | 0.0198 | 0.476 | 0.1090 |
| Partridge | 266.8 | 2/76 | 0.642 | 0.0640 | 0.3452 | 0.0217 | 0.465 | 0.1074 |
| Ostrich | 300 | 26/7 | 0.680 | 0.0569 | 0.3070 | 0.0229 | 0.458 | 0.1057 |
| Merlin | 336.4 | 18/1 | 0.684 | 0.0512 | 0.2767 | 0.0222 | 0.462 | 0.1055 |
| Linnet | 336.4 | 26/7 | 0.721 | 0.0507 | 0.2737 | 0.0243 | 0.451 | 0.1040 |
| Oriole | 336.4 | 30/7 | 0.741 | 0.0504 | 0.2719 | 0.0255 | 0.445 | 0.1032 |
| Chickadee | 397.5 | 18/1 | 0.743 | 0.0433 | 0.2342 | 0.0241 | 0.452 | 0.1031 |
| Ibis | 397.5 | 26/7 | 0.783 | 0.0430 | 0.2323 | 0.0264 | 0.441 | 0.1015 |
| Lark | 397.5 | 30/7 | 0.806 | 0.0427 | 0.2306 | 0.0277 | 0.435 | 0.1007 |
| Pelican | 477 | 18/1 | 0.814 | 0.0361 | 0.1947 | 0.0264 | 0.441 | 0.1004 |
| Flicker | 477 | 24/7 | 0.846 | 0.0359 | 0.1943 | 0.0284 | 0.432 | 0.0992 |
| Hawk | 477 | 26/7 | 0.858 | 0.0357 | 0.1931 | 0.0289 | 0.430 | 0.0988 |
| Hen | 477 | 30/7 | 0.883 | 0.0355 | 0.1919 | 0.0304 | 0.424 | 0.0980 |
| Osprey | 556.5 | 18/1 | 0.879 | 0.0309 | 0.1679 | 0.0284 | 0.432 | 0.0981 |
| Parakeet | 556.5 | 24/7 | 0.914 | 0.0308 | 0.1669 | 0.0306 | 0.423 | 0.0969 |
| Dove | 556.5 | 26/7 | 0.927 | 0.0307 | 0.1663 | 0.0314 | 0.420 | 0.0965 |
| Eagle | 556.5 | 30/7 | 0.953 | 0.0305 | 0.1651 | 0.0327 | 0.415 | 0.0957 |
| Peacock | 605 | 24/7 | 0.953 | 0.0283 | 0.1536 | 0.0319 | 0.418 | 0.0957 |
| Squab | 605 | 26/7 | 0.966 | 0.0282 | 0.1529 | 0.0327 | 0.415 | 0.0953 |
| Teal | 605 | 30/19 | 0.994 | 0.0280 | 0.1517 | 0.0341 | 0.410 | 0.0944 |
| Rook | 636 | 24/7 | 0.977 | 0.0269 | 0.1461 | 0.0327 | 0.415 | 0.0950 |
| Grosbeak | 636 | 26/7 | 0.990 | 0.0268 | 0.1454 | 0.0335 | 0.412 | 0.0946 |
| Egret | 636 | 30/19 | 1.019 | 0.0267 | 0.1447 | 0.0352 | 0.406 | 0.0937 |
| Flamingo | 666.6 | 24/7 | 1.000 | 0.0257 | 0.1397 | 0.0335 | 0.412 | 0.0943 |
| Crow | 715.5 | 54/7 | 1.051 | 0.0240 | 0.1304 | 0.0349 | 0.407 | 0.0932 |
| Starling | 715.5 | 26/7 | 1.081 | 0.0238 | 0.1294 | 0.0355 | 0.405 | 0.0948 |
| Redwing | 715.5 | 30/19 | 1.092 | 0.0237 | 0.1287 | 0.0373 | 0.399 | 0.0920 |

## *ACSR Data*

$$X_L = 2\pi fL = 2\pi f \times 2 \times 10^{-7} \ln\frac{GMD}{GMR} \qquad \Omega/m$$

$$X_L = 4\pi f \times 10^{-7} \ln\frac{GMD}{GMR} \qquad \Omega/m$$

$$X_L = 4\pi f \times 10^{-7} \times 1609 \times \ln\frac{GMD}{GMR} \qquad \Omega/mile$$

$$X_L = 2.022 \times 10^{-3} \times f \times \ln\frac{GMD}{GMR} \qquad \Omega/mile$$

$$X_L = \underbrace{2.022 \times 10^{-3} \times f \times \ln\frac{1}{GMR}}_{X_a} + \underbrace{2.022 \times 10^{-3} \times f \times \ln GMD}_{X_d} \qquad \Omega/mile$$

Term from table, depending on conductor type, but assuming a one foot spacing

Term independent of conductor, but with spacing $D_m$ in feet.

## ACSR Data

$$X_L = 2\pi fL = 2\pi f \times 2 \times 10^{-7} \ln \frac{GMD}{GMR} \qquad \Omega/m$$

$$X_L = 4\pi f \times 10^{-7} \ln \frac{GMD}{GMR} \qquad \Omega/m$$

$$X_L = 4\pi f \times 10^{-7} \times 1609 \times \ln \frac{GMD}{GMR} \qquad \Omega/mile$$

$$X_L = 2.022 \times 10^{-3} \times f \times \ln \frac{GMD}{GMR} \qquad \Omega/mile$$

$$X_L = \underbrace{2.022 \times 10^{-3} \times f \times \ln \frac{1}{GMR}}_{X_a} + \underbrace{2.022 \times 10^{-3} \times f \times \ln GMD}_{X_d} \qquad \Omega/mile$$

If both GMR and GMD are in feet,
then $X_a$ represents the inductive reactance at 1 ft spacing,
and $X_d$ is called the inductive reactance spacing factor.

To use the phase to neutral capacitance from table

$$X_C = \frac{1}{2\pi f C} \quad \Omega\text{-m} \quad \text{where } C = \frac{2\pi\varepsilon_0}{\ln \dfrac{D_m}{r}}$$

$$= \frac{1}{f} \times 1.779 \times 10^6 \ln \frac{D_m}{r} \quad \Omega\text{-mile (table is in } M\Omega\text{-mile)}$$

$$= \frac{1}{f} \times 1.779 \times \ln \frac{1}{r} + \frac{1}{f} \times 1.779 \times \ln D_m \quad M\Omega\text{-mile}$$

Term from table, depending on conductor type, but assuming a one foot spacing

Term independent of conductor, but with spacing $D_m$ in feet.

# Dove Example

$GMR = 0.0313$ feet

Outside Diameter = 0.07725 feet (radius = 0.03863)

Assuming a one foot spacing at 60 Hz

$$X_a = 2\pi 60 \times 2 \times 10^{-7} \times 1609 \times \ln \frac{1}{0.0313} \ \Omega/\text{mile}$$

$X_a = 0.420 \ \Omega/\text{mile}$, which matches the table

For the capacitance

$$X_C = \frac{1}{f} \times 1.779 \times 10^6 \ln \frac{1}{r} = 9.65 \times 10^4 \ \Omega\text{-mile}$$

# Example 1

Find the inductive reactance per mile and the capacitive reactance in MΩ.miles of a single phase line operating at 60 Hz. The conductor used is Partridge, with 20 ft spacing between the conductor centers.

D = 20 ft

# Example 1

Find the inductive reactance per mile and the capacitive reactance in MΩ.miles of a single phase line operating at 60 Hz. The conductor used is Partridge, with 20 ft spacing between the conductor centers.

○          ○
|← D = 20 ft →|

From the Tables, for Partridge conductor,
GMR = 0.0217 ft
and inductive reactance at 1 ft spacing
$X_a$ = 0.465 Ω/mile, which matches the table.
The spacing factor for 20 ft spacing is
$X_d$ = 0.3635 Ω /mile.
The inductance of the line is then
$X_L = X_a + X_d = 0.465 + 0.3635 = 0.8285$ Ω/ mile

# Example 1

Find the inductive reactance per mile and the capacitive reactance in MΩ.miles of a single phase line operating at 60 Hz. The conductor used is Partridge, with 20 ft spacing between the conductor centers.

○          ○
|← D = 20 ft →|

The outside radius of the Partridge conductor is $r = \dfrac{0.642}{2}$ in $= 0.0268$ ft

The capacitive reactance is

$$X_C = \frac{1.779 \times 10^6}{f} \ln\frac{D}{r} = \frac{1.779 \times 10^6}{f} \ln\frac{20}{0.0268} = 0.1961 \quad M\Omega. \text{mile}$$

# *Example 1*

Find the inductive reactance per mile and the capacitive reactance in MΩ.miles of a single phase line operating at 60 Hz. The conductor used is Partridge, with 20 ft spacing between the conductor centers.

○        ○

|←———  D = 20 ft  ———→|

The outside radius of the Partridge conductor is $r = \dfrac{0.642}{2}$ in = 0.0268 ft

The capacitive reactance is

$$X_C = \frac{1.779 \times 10^6}{f}\ln\frac{D}{r} = \frac{1.779 \times 10^6}{f}\ln\frac{20}{0.0268} = 0.1961 \quad M\Omega.\,mile$$

This is the capacitive reactance between the conductor and the neutral.

OR From tables $X_a' = 0.1074 \quad M\Omega.\,mile$

$X_d' = 0.0889 \quad M\Omega.\,mile$ for 20' spacing

$$\therefore X_C = X_a' + X_d' = 0.1963 \quad M\Omega.\,mile$$

---

# *Example 1*

Find the inductive reactance per mile and the capacitive reactance in MΩ.miles of a single phase line operating at 60 Hz. The conductor used is Partridge, with 20 ft spacing between the conductor centers.

○        ○

|←———  D = 20 ft  ———→|

The outside radius of the Partridge conductor is $r = \dfrac{0.642}{2}$ in = 0.0268 ft

The capacitive reactance is

$$X_C = \frac{1.779 \times 10^6}{f}\ln\frac{D}{r} = \frac{1.779 \times 10^6}{f}\ln\frac{20}{0.0268} = 0.1961 \quad M\Omega.\,mile$$

This is the capacitive reactance between the conductor and the neutral. Line-to-line capacitive reactance is:

OR From tables $X_a' = 0.1074 \quad M\Omega.\,mile$

$X_d' = 0.0889 \quad M\Omega.\,mile$ for 20' spacing

$$\therefore X_C = X_a' + X_d' = 0.1963 \quad M\Omega.\,mile$$

$$X_C^{L-L} = \frac{X_C}{2} = 0.0981 \quad M\Omega.\,mile$$

## Example 2

A three phase line operated at 60 Hz is arranged as shown. The conductors are ACSR Drake. If the length of the line is 175 miles and the normal operating voltage is 220 kV, Find:

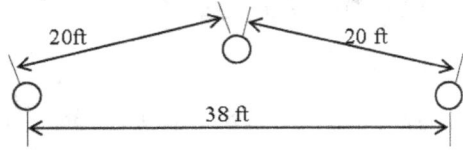

1. the inductive reactance per mile
2. the inductive reactance for the entire length of the line
3. the capacitive reactance for one mile
4. the capacitive reactance to neutral for the entire length of the line
5. the charging current for the line
6. the charging reactive power

---

## Example 2

### 1. The inductive reactance per mile:

For ACSR Drake conductor, GMR = 0.0373 ft

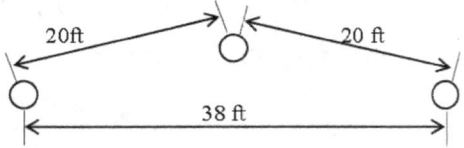

$$D_{eq} = \sqrt[3]{20 \times 20 \times 38} = 24.8 \quad ft$$

$$L = 2 \times 10^{-7} \ln\frac{24.8}{0.0373} = 13 \times 10^{-7} \qquad H/m$$

$$X_L = 2\pi \times 60 \times 1609 \times 13 \times 10^{-7} = 0.788 \qquad \Omega/mile$$

# *Example 2*

## 1. The inductive reactance per mile:

For ACSR Drake conductor, GMR = 0.0373 ft

$$D_{eq} = \sqrt[3]{20 \times 20 \times 38} = 24.8 \quad \text{ft}$$

$$L = 2 \times 10^{-7} \ln\frac{24.8}{0.0373} = 13 \times 10^{-7} \qquad H/m$$

$$X_L = 2\pi \times 60 \times 1609 \times 13 \times 10^{-7} = 0.788 \qquad \Omega/\text{mile}$$

**or,**

from the tables $X_a = 0.399 \quad \Omega/\text{mile}$

The spacing factor is calculated for spacing equal the geometric mean distance between the conductors, that is, $X_d = 2.022 \times 10^{-3} \times 60 \ln 24.8 = 0.389 \qquad \Omega/\text{mile}$

Then the line inductance is $X_{line} = X_a + X_d = 0.788 \quad \Omega/\text{mile}$ per phase

# *Example 2*

## 2. The inductive reactance the entire length of the line:

$X_L = 0.788 \times 175 = 137.9 \; \Omega$

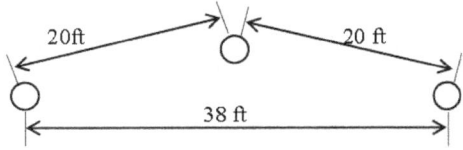

# Example 2

## 3. The capacitive reactance for one mile:

1 foot = 12 inch

The outside radius for Drake conductors is $r = \dfrac{1.108}{2}$ in $= 0.0462$ ft

The geometric mean distance for this line is

$$D_{eq} = \sqrt[3]{20 \times 20 \times 38} = 24.8 \quad \text{ft}$$

From tables, $X_a' = 0.0912 \quad M\Omega.\,\text{mile}$

$$X_d' = \frac{1.779 \times 10^6}{f}\ln D_{eq} = \frac{1.779 \times 10^6}{60}\ln 24.8 = 0.0952 \quad M\Omega.\,\text{mile}$$

$$\therefore X_{cn} = X_a' + X_d' = 0.1864 \quad M\Omega.\,\text{mile}$$

This is the capacitive reactance to neutral.

# Example 2

## 4. The capacitive reactance to neutral for the entire length of the line:

For the length of 175 miles,

$$X_{Ctotal} = \frac{X_{cn}}{175} = 1065 \quad \Omega$$

## *Example 2*

5. the charging current for the line:

$$I_C = \frac{V_{LN}}{X_{Ctotal}} = \frac{\frac{220k}{\sqrt{3}}}{1065} = 119 \quad A$$

6. the charging reactive power:

$$Q_C = \sqrt{3}V_{LL}I_C = \sqrt{3} \times 220k \times 119 = 45.45 \quad MVAr$$

# EEL 2023

# Power Generation and Transmission

## Chapter 7
### Transmission Line Equivalent Circuit

# Transmission line models

➢ Transmission lines are characterized by their distributed parameters: distributed resistance, inductance, and capacitance.

➢ The distributed series and shunt elements of the transmission line make it harder to model. Such parameters may be approximated by many small discrete resistors, capacitors, and inductors.

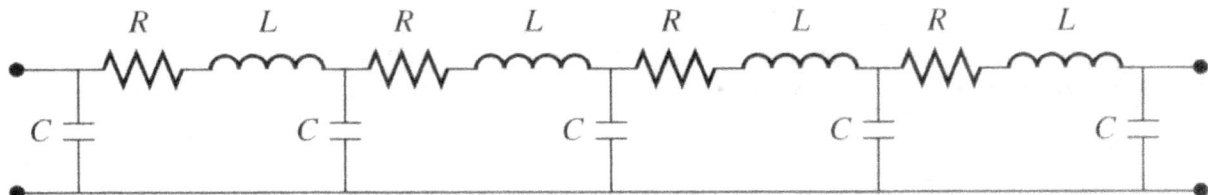

➢ However, this approach is not very practical, since it would require to solve for voltages and currents at all nodes along the line. We could also solve the exact differential equations for a line but this is also not very practical for large power systems with many lines.

2

Fortunately, certain simplifications can be used...

➢ Overhead transmission lines shorter than 80 km (50 miles) can be modeled as a series resistance and inductance, since the shunt capacitance can be neglected over short distances.

➢ The inductive reactance for OHTLs is typically much larger than the resistance of the line.

➢ For medium-length lines (80-240 km), shunt capacitance should be taken into account. However, it can be modeled by two capacitors of a half of the line capacitance each.

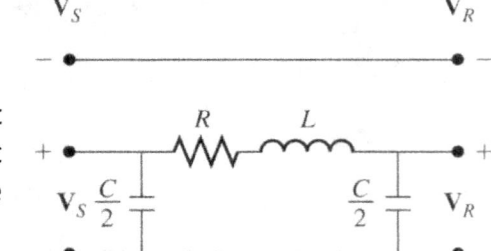

➢ Lines longer than 240 km (150 miles) are long transmission lines and are to be discussed later.

➤ The total series resistance, series reactance, and shunt admittance of a transmission line can be calculated as

$$R = rd \tag{1}$$

$$X = xd \tag{2}$$

$$Y = yd \tag{3}$$

➤ where *r*, *x*, and *y* are resistance, reactance, and shunt admittance per unit length and *d* is the length of the transmission line. The values of *r*, *x*, and *y* can be computed from the line geometry or found in the reference tables for the specific transmission line.

## Short transmission line

The per-phase equivalent circuit of a short line

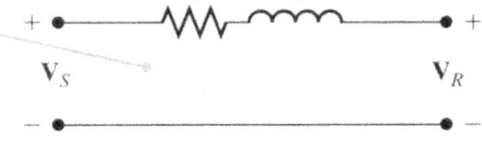

➤ $V_S$ and $V_R$ are the sending and receiving end voltages; $I_S$ and $I_R$ are the sending and receiving end currents. Assumption of no shunt admittance leads to

$$I_S = I_R \tag{4}$$

➤ We can relate voltages through the Kirchhoff's voltage law

$$V_S = V_R + ZI = V_R + RI + jX_L I \tag{5}$$

$$V_R = V_S - RI - jX_L I \tag{6}$$

➤ which is very similar to the equation derived for a synchronous generator.

## Short transmission line: phasor diagram

AC voltages are usually expressed as phasors.

1- Load with lagging power factor:

2- Load with unity power factor:

3- Load with leading power factor:

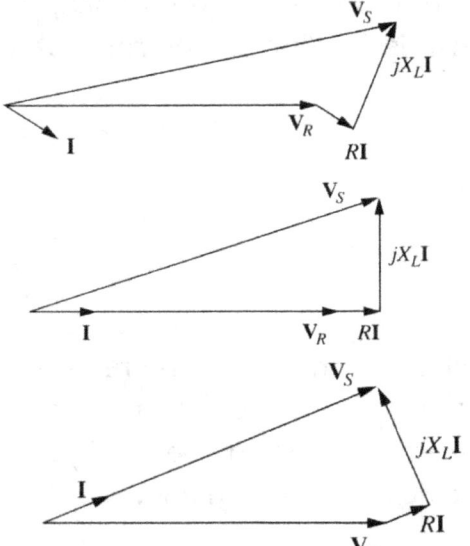

For a given source voltage $V_S$ and magnitude of the line current, the received voltage is lower for lagging loads **and higher for leading loads.**

# Transmission line characteristics

➢ In real overhead transmission lines, the line reactance $X_L$ is normally much larger than the line resistance $R$; therefore, the line resistance is often neglected. We consider next some important transmission line characteristics...

➢ Assuming that a single generator supplies a single load through a transmission line, we consider consequences of increasing load

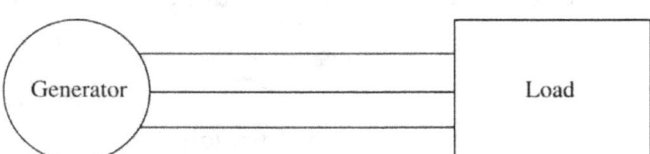

➢ Assuming that the generator is ideal, an increase of load will increase a real and (or) reactive power drawn from the generator and, therefore, increase the line current, while the voltage will be unchanged.

1) If more load is added with the same lagging power factor, the magnitude of the line current increases but the current remains at the same angle $\theta$ with respect to $V_R$ as before.

➤ The voltage drop across the reactance increases but stays at the same angle.

➤ Assuming zero line resistance and remembering that the source voltage has a constant magnitude:

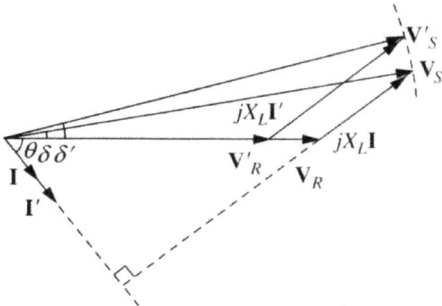

$$V_S = V_R + jX_L I \qquad (7)$$

➤ voltage drop across reactance $jX_L I$ will stretch between $V_R$ and $V_S$.

Therefore, when a lagging load increases, the received voltage decreases sharply.

2) **An** increase in a unity PF load on the other hand, slightly decrease the received voltage at the end of the transmission line.

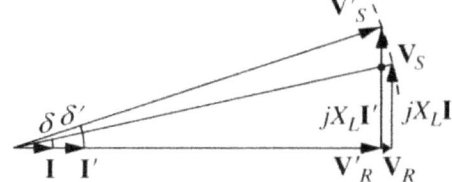

3) **Finally, an** increase in a load with leading PF increases the received (terminal) voltage of the transmission line.

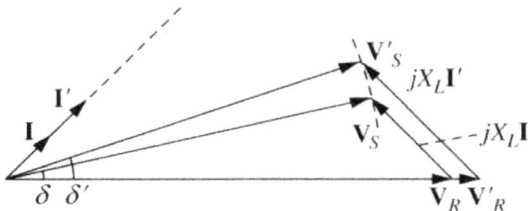

In summary:

1. If lagging (inductive) loads are added at the end of a line, the voltage at the end of the transmission line decreases significantly – large positive VR.
2. If unity-PF (resistive) loads are added at the end of a line, the voltage at the end of the transmission line decreases slightly – small positive VR.
3. If leading (capacitive) loads are added at the end of a line, the voltage at the end of the transmission line increases – negative VR.

The voltage regulation of a transmission line is

$$VR = \frac{V_{nl} - V_{fl}}{V_{fl}} \cdot 100\% \qquad (8)$$

where $V_{nl}$ and $V_{fl}$ are the no-load and full-load voltages at the line output.

## Power flow in a transmission line

The real power input to a 3-phase transmission line can be computed as

$$P_{in} = 3V_S I_S \cos\theta_S = \sqrt{3}V_{LL,S} I_S \cos\theta_S \qquad (9)$$

where $V_S$ is the magnitude of the source (input) line-to-neutral voltage and $V_{LL,S}$ is the magnitude of the source (input) line-to-line voltage. Note that Y-connection is assumed! Similarly, the real output power from the transmission line is

$$P_{out} = 3V_R I_R \cos\theta_R = \sqrt{3}V_{LL,R} I_R \cos\theta_R \qquad (10)$$

The reactive power input to a 3-phase transmission line can be computed as

$$Q_{in} = 3V_S I_S \sin\theta_S = \sqrt{3}V_{LL,S} I_S \sin\theta_S \qquad (11)$$

And the reactive output power is

$$Q_{out} = 3V_R I_R \sin\theta_R = \sqrt{3}V_{LL,R} I_R \sin\theta_R \qquad (12)$$

The apparent power input to a 3-phase transmission line can be computed as

$$S_{in} = 3V_S I_S = \sqrt{3}V_{LL,S} I_S \qquad (13)$$

And the apparent output power is

$$S_{out} = 3V_R I_R = \sqrt{3}V_{LL,R} I_R \qquad (14)$$

➤ If the resistance $R$ is ignored, the output power of the transmission line can be simplified…

➤ A simplified phasor diagram of a transmission line indicating that $I_S = I_R = I$.
We further observe that the vertical segment $bc$ can be expressed as either $V_S \sin\delta$ or $X_L I\cos\theta$. Therefore:

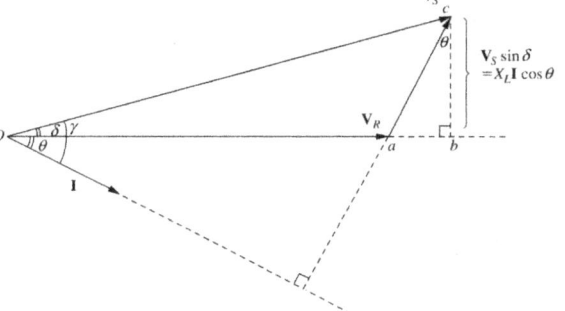

$$I\cos\theta = \frac{V_S \sin\delta}{X_L} \qquad (15)$$

➤ Then the output power of the transmission line equals to its input power (because $R$ is ignored):

$$P = \frac{3V_S V_R \sin\delta}{X_L} \qquad (16)$$

➤ Therefore, the power supplied by a transmission line depends on the angle between the phasors representing the input and output voltages.

➤ The maximum power supplied by the transmission line occurs when $\delta = 90^0$:

$$P_{max} = \frac{3V_S V_R}{X_L} \qquad (17)$$

➤ This maximum power is called the steady-state stability limit of the transmission line. The real transmission lines have non-zero resistance and, therefore, overheat long before this point. Full-load angles of $25^0$ are more typical for real transmission lines.

➤ Few interesting observations can be made from the power expressions:

1. The maximum power handling capability of a transmission line is a function of the *square of its voltage.* For instance, if all other parameters are equal, a 220 kV line will have 4 times the power handling capability of a 110 kV transmission line.

➤ Therefore, it is beneficial to increase the voltage… However, very high voltages produce very strong electromagnetic fields (interferences) and may produce a corona – glowing of ionized air that substantially increases losses.

2. The maximum power handling capability of a transmission line is inversely proportional to its series reactance, which may be a serious problem for long transmission lines. Some very long lines include **series capacitors** to reduce the total series reactance and thus increase the total power handling capability of the line.

3. In a normal operation of a power system, the magnitudes of voltages $V_S$ and $V_R$ do not change much, therefore, the angle $\delta$ basically controls the power flowing through the line. It is possible to control power flow by placing a phase-shifting transformer at one end of the line and varying voltage phase.

## Transmission line efficiency

The efficiency of the transmission line is

$$\eta = \frac{P_{out}}{P_{in}} \cdot 100\%$$

(18)

## 4. Transmission line ratings

One of the main limiting factors in transmission line operation is its resistive heating. Since this heating is a function of the square of the current flowing through the line and does not depend on its phase angle, transmission lines are typically rated at a nominal voltage and apparent power.

## Transmission line limits

Several practical constrains limit the maximum real and reactive power that a transmission line can supply. The most important constrains are:

1. The maximum steady-state current must be limited to prevent the overheating in the transmission line. The power lost in a line is approximated as

$$P_{loss} = 3I_L^2 R \qquad (19)$$

The greater the current flow, the greater the resistive heating losses.

2. The voltage drop in a practical line should be limited to approximately 5%. In other words, the ratio of the magnitude of the receiving end voltage to the magnitude of the sending end voltage should be

$$\frac{|VR|}{|Vs|} \geq 0.95 \qquad (20)$$

This limit prevents excessive voltage variations in a power system.

3. The angle $\delta$ in a transmission line should typically be $\leq 30^0$ ensuring that the power flow in the transmission line is well below the static stability limit and, therefore, the power system can handle transients.

Any of these limits can be more or less important in different circumstances:

In short lines, where series reactance $X$ is relatively small, the resistive heating usually limits the power that the line can supply.

In longer lines operating at **lagging** power factors, the voltage drop across the line is usually the limiting factor.

In longer lines operating at **leading** power factors, the maximum angle $\delta$ can be the limiting factor.

---

# Two-port networks and ABCD models

➤ A transmission line can be represented by a 2-port network – a network that can be isolated from the outside world by two connections (ports) as shown.

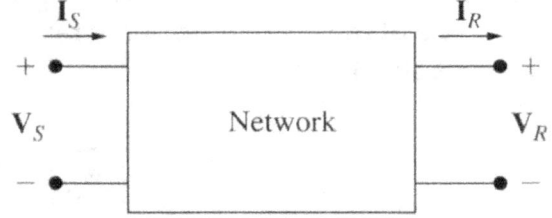

➤ If the network is linear, an elementary circuits theorem (analogous to Thevenin's theorem) establishes the relationship between the sending and receiving end voltages and currents as

$$V_S = AV_R + BI_R$$
$$I_S = CV_R + DI_R$$

(21)

➤ Here constants $A$ and $D$ are dimensionless, a constant $B$ has units of $\Omega$, and a constant $C$ is measured in siemens. These constants are sometimes referred to as generalized circuit constants, **or** ABCD constants.

$$V_S = AV_R + BI_R$$
$$I_S = CV_R + DI_R$$

(21)

➤ The ABCD constants can be physically interpreted. Constant *A* represents the effect of a change in the receiving end voltage on the sending end voltage; and constant *D* models the effect of a change in the receiving end current on the sending end current. Naturally, both constants *A* and *D* are dimensionless.

➤ The constant *B* represents the effect of a change in the receiving end current on the sending end voltage. The constant *C* denotes the effect of a change in the receiving end voltage on the sending end current.

➤ Transmission lines are 2-port linear networks, and they are often represented by ABCD models. For the short transmission line model, $I_S = I_R = I$, and the ABCD constants are

$$A = 1$$
$$B = Z$$
$$C = 0$$
$$D = 1$$

(22)

# *Medium-length transmission line*

➤ Considering medium-length lines (50 to 150 mile-long), the shunt admittance must be included in calculations. However, the total admittance is usually modeled ($\pi$ model) as two capacitors of equal values (each corresponding to a half of total admittance) placed at the sending and receiving ends.

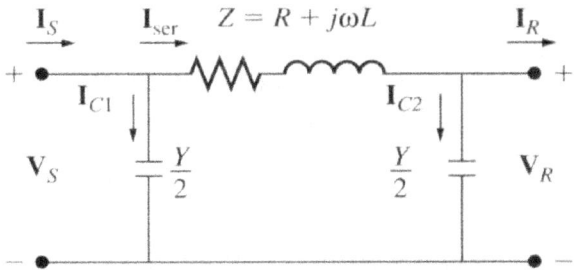

➤ The current through the receiving end capacitor can be found as

$$I_{C2} = V_R \frac{Y}{2}$$

(23)

➤ And the current through the series impedance elements is

$$I_{ser} = V_R \frac{Y}{2} + I_R$$

(24)

➤ From the Kirchhoff's voltage law, the sending end voltage is

$$V_S = ZI_{ser} + V_R = Z\left(I_{C2} + I_R\right) + V_R = \left(\frac{YZ}{2} + 1\right)V_R + ZI_R \qquad (25)$$

The source current will be

$$I_S = I_{C1} + I_{ser} = I_{C1} + I_{C2} + I_R = V_S\frac{Y}{2} + V_R\frac{Y}{2} + I_R = Y\left(\frac{ZY}{4} + 1\right)V_R + \left(\frac{ZY}{2} + 1\right)I_R \qquad (26)$$

Therefore, the ABCD constants of a medium-length transmission line are

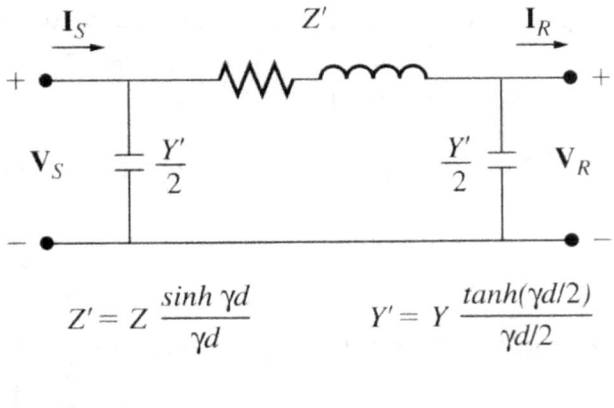

$$A = \frac{ZY}{2} + 1$$

$$B = Z$$

$$C = Y\left(\frac{ZY}{4} + 1\right) \qquad (27)$$

$$D = \frac{ZY}{2} + 1$$

➤ If the shunt capacitance of the line is ignored, the ABCD constants are the constants for a short transmission line.

# Long transmission line

➤ For long lines, it is not accurate enough to approximate the shunt admittance by two constant capacitors at either end of the line. Instead, both the shunt capacitance and the series impedance must be treated as distributed quantities; the voltages and currents on the line should be found by solving differential equations of the line.

However, it is possible to model a long transmission line as a $\pi$ model with a *modified* series impedance $Z'$ and a *modified* shunt admittance $Y'$ and to perform calculations on that model using ABCD constants. The modified values of series impedance and shunt admittance are:

$$Z' = Z\frac{\sinh \gamma d}{\gamma d}$$

$$Y' = Y\frac{\tanh\left(\gamma d / 2\right)}{\gamma d / 2} \qquad (28)$$

$$Z' = Z\frac{\sinh \gamma d}{\gamma d} \qquad Y' = Y\frac{\tanh(\gamma d/2)}{\gamma d/2}$$

➢ Here $Z$ is the series impedance of the line; $Y$ is the shunt admittance of the line; $d$ is the length of the line; $\gamma$ is the propagation constant of the line:

$$\gamma = \sqrt{yz} \qquad (29)$$

➢ where $y$ is the shunt admittance per kilometer and $z$ is the series impedance per km.

➢ As $\gamma d$ gets small, the ratios approach 1.0 and the model becomes a medium-length line model. The ABCD constants for a long transmission line are

$$
\begin{aligned}
A &= \frac{Z'Y'}{2} + 1 \\
B &= Z' \\
C &= Y'\left(\frac{Z'Y'}{4} + 1\right) \\
D &= \frac{Z'Y'}{2} + 1
\end{aligned}
\qquad (30)
$$

24

# Example 1 – Short Lines

- A *12 Km* long three phase overhead line delivers *7.5 MW* at 50 Hz *33 kV* at a power factor of *0.78* lagging. Line loss is *13.5 %* of the power delivered. Line inductance is *1.2 mH* per *km* per phase. Calculate the sending end voltage and line regulation.

# Example 1

- Receiving end voltage, $V_r = 33000/\sqrt{3} = 19053\ V$

- Now $\quad Power\ Factor, PF = \dfrac{Real\ Power}{Apparent\ Power} = \dfrac{P}{S} = \dfrac{P}{3E_r I_r}$

- Line current, $\quad I_r = \dfrac{P}{\sqrt{3} \times E_{line} \times PF} = \dfrac{7.5 \times 10^6}{\sqrt{3} \times 33000 \times 0.78} = 175\ A$

- Total line loss in conductors, $\quad \dfrac{13.5}{100} \times 7.5 \times 10^6 = 1.0125 \times 10^6\ W$

- So, $\quad 3 \times I^2 R = 1.0125 \times 10^6$

- Or, $\quad R = \dfrac{1.0125 \times 10^6}{3 \times 175 \times 175} = 11.02\Omega$

- Reactance, $\quad X = 2\pi \times 50 \times 1.2 \times 10^{-3} \times 12 = 4.52\Omega$


# Example 1 (continued)

- Sending end voltage, $\quad V_S \approx V_r + I_r R cos\emptyset_r + I_r X sin\emptyset_r$

  $= 19053 + 175 \times 11.03 \times 0.78 + 175 \times 4.52 \times 0.63 = 21.053\ kV$

- Line-to-line, sending end voltage $\quad = \dfrac{\sqrt{3} \times 21053}{1000} = 36.464\ kV$

- and regulation $\quad = \dfrac{|V_S| - |V_r|}{|V_r|} = \dfrac{21053 - 19053}{19053} = 0.105 = 10.5\%$

# Example 2

A single-phase transmission line having an inductive reactance of 7.5 Ω is connected to a fixed sender voltage of 2.2 kV

Calculate

a) The maximum active power the line can deliver to a resistive load

b) The corresponding receiver voltage

c) The receiver power when the receiver voltage is 2090 V.

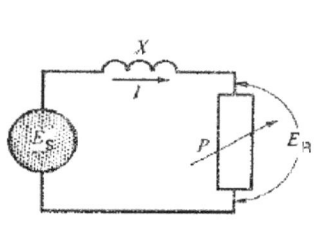

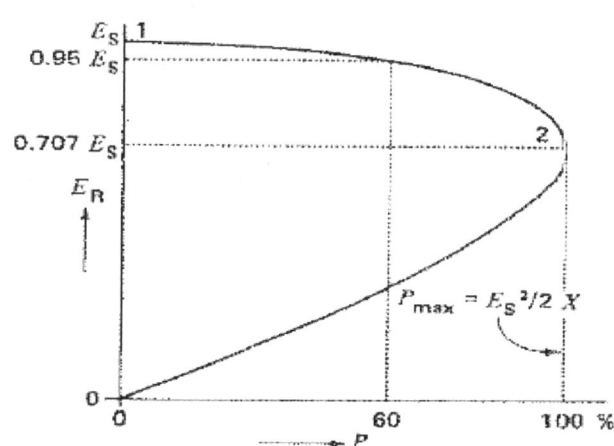

- **Ans.**
- **a)   The maximum power that can be transmitted to the load is**

$$P_{max} = E_S^2/2X \quad = \frac{2200^2}{2 \times 7.5} = 322.67 \, kW$$

- **b)   The corresponding receiver voltage is**
- **c)   In order to calculate the receiver power when Er = 2090 V, we first calculate the value of the current I.**
- **Taking Er = 2090 as the reference phasor, we draw the phasor diagram for the circuit.**     $E_r = 0.707E_s = 0.707 \times 2200 = 1555.4V$

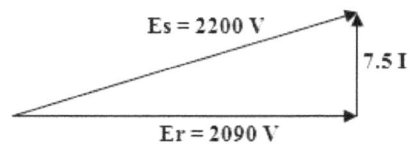

- **Current I is in phase with Er because the load is resistive. We can write,**
- **Es = Er + jIX   = 2090 + 7.5jI**

From the phasor diagram we can apply Pythagoras and write,

$$E_s^2 = E_r^2 + (7.5I)^2$$
$$2200^2 = 2090^2 + 57.75I^2$$

From this we have I = $\sqrt{8171.4}$ = 90.40 A

The power to the receiver is, therefore,

$\quad$ P = Es I = 2090 x 90.40 = 188.94 kW

Note that 188.94 kW is almost 60% of Pmax (322.67 kW) as predicted from the graph of fig. 2